USING RECLAIMED WATER TO AUGMENT POTABLE WATER RESOURCES

A Special Publication

Second Edition

D1428426

2008

Water Environment Federation
601 Wythe Street
Alexandria, VA 22314-1994 USA

American Water Works Association
6666 West Quincy Avenue
Denver, CO 80235 USA

Library of Congress Cataloging-in-Publication Data

Using reclaimed water to augment potable water resources : a special
publication.—2nd ed.
   p. cm.
Includes bibliographical references and index.
ISBN 978-1-57278-250-1
1. Water reuse.   2. Drinking water.   3. Artificial groundwater recharge.
I. Water Environment Federation.   II. American Water Works Association.
TD429.U85 2008
363.6'1—dc22

                                                         2008033224

Water Environment Research, WEF, and WEFTEC are registered trademarks of the
Water Environment Federation.

IMPORTANT NOTICE

The material presented in this publication has been prepared in accordance with generally recognized engineering principles and practices and is for general information only. This information should not be used without first securing competent advice with respect to its suitability for any general or specific application.

The contents of this publication are not intended to be a standard of the Water Environment Federation (WEF) or American Water Works Association (AWWA) and are not intended for use as a reference in purchase specifications, contracts, regulations, statutes, or any other legal document.

No reference made in this publication to any specific method, product, process, or service constitutes or implies an endorsement, recommendation, or warranty thereof by WEF or AWWA.

WEF and AWWA make no representation or warranty of any kind, whether expressed or implied, concerning the accuracy, products, or processes discussed in this publication and assumes no liability.

Anyone using this information assumes all liability arising from such use, including but not limited to infringement of any patent or patents.

Prepared by a Joint Using Reclaimed Water to Augment Potable Water Resources Task Force of the Water Environment Federation and the American Water Works Association

Kevin D. Conway, P.E., *Chair*

David K. Ammerman, P.E.

Robert A. Bergman

Barry Bohn, P.E.

Guy W. Carpenter, P.E.

James Crook, Ph.D., P.E.

Christine DeBarbadillo

Jörg E. Drewes, Ph.D.

Val S. Frenkel, Ph.D., P.Eng.

Mikel E. Goldblatt

Gary Grinnell, P.E.

Blanca E. Jimenez Cisneros

Gustavo V. Lopez, P.E.

Linda Macpherson

Mark Mason, P.E.

Robert L. Matthews

Margaret H. Nellor, P.E.

Marie-Laure Pellegrin, Ph.D.

John P. Rehring. P.E., BCEE

Joseph C. Reichenberger, P.E., BCEE

Andrew W. Richardson, P.E., BCEE

Alan Rimer

Eric Rosenblum

Robert Rubin, Ph.D.

John Ruetten

Donald Safrit

Roger E. Schenk

Larry Schimmoller, P.E.

Jeffrey Stone

Peter J. H. Thomson, P.E.

Cindy Wallis-Lage

David W. York, Ph.D., P.E.

Under the direction of the WEF Watershed Subcommittee of the Technical Practice Committee and the AWWA Water Reuse Committee of the Water Conservation Division

About the Water Environment Federation

Formed in 1928, the Water Environment Federation (WEF) is a not-for-profit technical and educational organization with 35,000 individual members and 81 affiliated Member Associations representing an additional 50,000 water quality professionals throughout the world. WEF and its member associations proudly work to achieve our mission of preserving and enhancing the global water environment.

For information on membership, publications, and conferences, contact

Water Environment Federation
601 Wythe Street
Alexandria, VA 22314-1994 USA
(703) 684-2400
http://www.wef.org

About the American Water Works Association

Founded in 1881, American Water Works Association (AWWA) is an international nonprofit and educational society with more than 60,000 members. AWWA advances public health, safety and welfare by uniting the efforts of the full spectrum of the water community. Membership includes more than 4,600 utilities that supply drinking water to roughly 180 million people in North America. Through our collective strength we become better stewards of water for the greatest good of the people and the environment.

For information on membership, publications, and conferences, contact

American Water Works Association
6666 West Quincy Avenue
Denver, CO 80235 USA
1-(303)-794-7711

Or visit our Web site at http://www.awwa.org

Contents

List of Tables

List of Figures

Special Publications of the Water Environment Federation

The WEF Technical Practice Committee (formerly the Committee on Sewage and Industrial Wastes Practice of the Federation of Sewage and Industrial Wastes Associations) was created by the Federation Board of Control on October 11, 1941. The primary function of the Committee is to originate and produce, through appropriate subcommittees, special publications dealing with technical aspects of the broad interests of the Federation. These publications are intended to provide background information through a review of technical practices and detailed procedures that research and experience have shown to be functional and practical.

Water Environment Federation Technical Practice Committee Control Group

B. G. Jones, *Chair*
J. A. Brown, *Vice-Chair*

Preface

The second edition of this Special Publication was produced under the direction of Kevin D. Conway, P.E., Chair. Principal authors of the publication and its respective chapters are

Guy Carpenter, P.E.	(2)
Kevin D. Conway, P.E.	(1)
Margaret H. Nellor, P.E.	(3)
Joseph Reichenberger, P.E., BCEE	(5)
Andy Richardson, P.E., BCEE	(4)
John Ruetten	(6)

Nathan S. Lester contributed to the development of Chapter 4.

The purpose of this updated publication is to provide a compilation of technical information on indirect potable reuse for water reuse technical practitioners. Since publishing *Using Reclaimed Water To Augment Potable Water Resources* in 1998, several practices have advanced that support an update of this special publication. These practices include the continuing evolution of regulations across the United States, the continuing advancement of technologies that provide proper barriers, and utilities' ongoing, full-scale, indirect potable reuse experiences. These advancements continue to provide the public with a safe and reliable water resource option.

When considering indirect potable reuse options, water resource planners also should fully evaluate nonpotable reuse alternatives, if possible. The intention of this publication is to assist water resource planners who wish to evaluate both indirect potable and nonpotable reuse options.

This publication is valuable to the water-management industry; it was not prepared as a communications tool for the public. The technical writing in this publication is in the language of engineers, scientists, and planners.

Authors' and reviewers' efforts were supported by the following organizations:

Black & Veatch, Cary, North Carolina, and Kansas City, Missouri
Boyle Engineering Corporation, Orlando, Florida
CDM, Austin, Texas; Denver, Colorado; Maitland, Florida; and Rancho
    Cucamonga, California
Chester Engineers, Inc., Potomac, Maryland

CH2M HILL, Englewood, Colorado; Gainesville, Florida; and Portland, Oregon
City of San Jose Department of Environmental Services, San Jose, California
Greeley and Hansen, Phoenix, Arizona
Colorado School of Mines, Golden, Colorado
HDR Engineering, Inc., Phoenix, Arizona
Hydroqual, Inc., Mahwah, New Jersey
Idaho Department of Environmental Quality, Boise, Idaho
Insto de Ingenieria, UNAM, Hidraulica Y Ambiental, Coyoacan DF Mexico
Kennedy/Jenks Consultants, San Francisco, California
Las Vegas Valley Water District, Las Vegas, Nevada
Loyola Marymount University, Los Angeles, California
Nellor Environmental Associates, Inc., Austin, Texas
North Carolina State University, Department of Biological Agriculture
    Engineering, Raleigh, North Carolina
Resource Trends, Inc., Escondido, California
Watertech Engineering Research & Health, Inc., Calgary, Alberta, Canada
York Water Circle, Tallahassee, Florida

Chapter 1

Introduction

Properly treated wastewater effluent (reclaimed water) has been used to meet nonpotable water needs in many parts of the United States for decades. Reclaimed water has been successfully used for a wide range of nonpotable uses, including landscape and agricultural irrigation, industrial process water, power plant cooling water, toilet flushing, car washing, augmentation of recreational waterbodies, fire protection, commercial cleansing, construction, and habitat restoration. The specific requirements for implementing these systems vary from state to state, but in general, using reclaimed water to meet nonpotable water demands is an accepted practice.

More recently, there has been an increasing interest in the use of highly treated reclaimed water to augment potable water resources. A few projects have been implemented, and several others are being planned. In these cases, water resources were limited, and planners compared the potential benefits of recycling reclaimed water with the potential disadvantages and concluded that the benefits outweighed the disadvantages. However, important questions remain about the levels of treatment, monitoring, and testing needed to ensure the safety of recycling "reclaimed water."

When discussing potable water reuse, it is important to distinguish between "direct" and "indirect" potable water reuse. Conceptually, direct potable reuse is a "pipe-to-pipe" connection between the reclaimed water treatment facility and the potable water distribution system. In indirect potable reuse, highly treated reclaimed water is introduced to a surface water or groundwater system that ultimately uses the water as a potable water supply. In an indirect system, reclaimed water is blended with water in the natural system, and there may be a significant delay and more treatment between the points where reclaimed water is discharged and where water is withdrawn into the potable water treatment facility. Reclaimed water is significantly diluted by natural water.

This publication provides a compilation of available information on this subject. When considering indirect potable reuse, water resource planners should fully evaluate nonpotable water reuse alternatives if possible. This publication is intended to help water resource planners evaluate both options.

1.0 CHAPTER 2—PLANNING INDIRECT POTABLE REUSE

Planned reuse was born out of a desire for an alternative to discharge and has evolved into a practice necessary to ensure that the water resources portfolio is robust. Indirect potable reuse (planned or unplanned) occurs when some fraction of the drinking water source has been previously used.

Unplanned indirect potable reuse is a common situation in watersheds or utility service areas without an integrated water-resources plan. Planned indirect potable reuse occurs when treated wastewater is intentionally introduced into the water supply after the affected agencies have considered the links between reclaimed water and potable water supply. Such planning involves the purposeful consideration of many interrelated technical, economic, public policy, and regulatory issues.

1.1   Integrated Water Resources Management

Increasingly within the United States, water-related services need to be provided within the context of an integrated water-resources plan because there are few instances in which water, wastewater, reclaimed water, and stormwater utilities do not affect one another, the environment, or water-related recreations.

Reclaimed water typically is reliable in terms of quality, quantity, and availability, so it is an important resource to consider when augmenting potable water resources. This option presumably would be considered because existing sources of drinking water supply are not available in sufficient quantity, quality, or reliability. Other approaches to consider include improved water delivery and use, nonpotable reuse, exchange of reclaimed water for other drinking water sources, and acquisition and development of new sources of supply.

There are numerous examples of successful and unsuccessful indirect potable reuse projects. The failures are largely attributed to negative branding because detractors perpetuated negative perceptions and fear of a source of water once considered a "waste." Most successful indirect potable reuse projects in the United States introduced a natural system between the water reclamation facility and the drinking water production facility. They also demonstrated the benefits of indirect potable reuse.

1.2   Means of Introducing Reclaimed Water to Water Supplies

Reclaimed water may be introduced to water supplies via groundwater recharge or surface water augmentation.

Groundwater aquifers can be used to store water (i.e., act as underground reservoirs) without the evaporative losses associated with surface water reservoirs. Moreover, the aquifer itself can function as a water conveyance system, providing lateral movement without pipelines. Depending on site-specific conditions, this method of storage and delivery may be significantly less expensive (in both capital and operational costs) and more desirable than a traditional aboveground storage and distribution facility. Water-quality degradation can be a concern for water held in either surface or aquifer storage systems. Generally, surface water sources are more readily cleansed of contamination than aquifers, but aquifers are less susceptible to rapid and extensive contamination.

Water can be introduced to groundwater aquifers via injection wells, recharge basins, or alluvial infiltration. There are three types of injection wells: direct-injection wells, aquifer storage and recovery (ASR) wells, and vadose zone recharge wells.

Like direct-injection wells, ASR wells deliver reclaimed water directly to the aquifer, but they also can recover water from the aquifer using the same column. Vadose zone recharge wells inject reclaimed water into the unsaturated zone between the ground surface and the water table of the saturated aquifer.

Recharge basins are shallow basins that receive reclaimed water and allow it to percolate down to the water table. Alluvial infiltration involves releasing water into a stream channel and allowing it to percolate to the water table.

Many factors affect the choice of method for a particular site and recharge program. These include land availability, topography, climatic factors, regulatory and permitting constraints, subsurface geology, reclaimed water quantities and quality, cost of facilities, environmental effects, and public perception.

Planned indirect water reuse via surface water augmentation is a practice that is gaining favor among water purveyors. It must be done carefully, in a manner appropriate for the distance between the delivery and intake points, the volume of water to be reused, and the watershed characteristics.

1.3 Institutional Considerations

Water rights issues can impose restrictions and requirements on reuse projects. Water rights doctrines have been established by state legislatures and courts in response to the state's hydrologic conditions. Federal doctrines are important in the body of water rights laws, but to a large extent, state doctrines form the fundamental core of water law.

1.4 Economics

The issues that can affect an indirect potable reuse project's financial feasibility are principally the water-quality goals or standards that must be achieved, the costs of alternatives, the financing method for capital costs, and the method for reimbursing operations and maintenance (O&M) costs.

2.0 CHAPTER 3—HEALTH AND REGULATORY CONSIDERATIONS

This chapter presents an overview of the health and regulatory issues that should be considered when planning and implementing an indirect potable reuse project. The perceived and actual risks of using reclaimed water for indirect potable projects are higher than those for nonpotable reuse because indirect reuse involves supplementing drinking water with reclaimed water. Potable reuse projects also can face challenges related to the water-quality or environmental concerns associated with co-mingling reclaimed water with surface water or groundwater.

Despite the prevalence of unplanned indirect potable reuse, some experts believe that water originating from raw wastewater is inherently more risky— even if the reclaimed water meets drinking water standards. Their concerns include the following:

- Disinfection of reclaimed water may create different (often unidentified) byproducts than are found in conventional water supplies;
- Only a small percentage of the organic compounds in drinking water have been identified, and the health effects of only a few of the identified compounds have been determined; and
- The health effects of mixtures of two or more of the hundreds of compounds in reclaimed water are not easily characterized.

While these issues are not insurmountable, they do illustrate the need to acknowledge that there are complex health and regulatory issues that must be evaluated and accommodated in indirect potable reuse projects, from public health, water-quality, and environmental-protection perspectives. Many of these concerns are common to drinking water.

2.1 Health Issues—Microorganisms

Hundreds of different types of pathogens excreted in the fecal material of infected hosts find their way into municipal wastewater. Diseases caused by waterborne microorganisms range from mild gastroenteritis to severe illnesses (e.g., infectious hepatitis, cholera, typhoid fever, and meningitis). For infection to occur, the infectious agent must be present in the community; it must survive all of the wastewater treatment processes to which it is exposed; an individual must directly or indirectly come into contact with reclaimed water; and the water must contain enough of the agent to cause infection at the time of contact.

In indirect potable reuse projects, the principal infectious agents of concern are bacteria, parasites, and viruses. The occurrence and concentration of pathogens in a particular wastewater are impractical to predict with any degree of confidence because of the numerous factors involved (e.g., sources, effectiveness of wastewater treatment, the health of the population in the collection area, the ability of infectious agents to survive outside their hosts in various environmental conditions, and the ability to measure specific pathogens). While raw wastewater may pose a significant risk to public health, current treatment technologies allow for significant pathogen reduction. Indirect potable reuse projects involving soil–aquifer treatment can remove even more pathogens (e.g., viruses and parasites) from the reclaimed water after it has left the treatment plant.

2.2   Chemical Constituents

All water supplies could be contaminated by chemical substances (e.g., organics, inorganics, and radionuclides). Municipal wastewater's possible constituents may be a concern when reclaimed wastewater is designated as a source of potable water. Part of this concern has to do with the components of reclaimed water that have known health-related properties and traditionally are monitored and regulated in drinking water. However, concerns also arise as a result of speculation about the presence and effects of unidentified chemicals—particularly organics. While such concerns may exist for any water supply, more typically are raised about reclaimed water because of its ostensible connection with sources of pollution.

With recent analytical advances and related reconnaissance studies of surface water and drinking water, microconstituents (e.g., endocrine-disrupting chemicals, pharmaceuticals, and personal care products) have been detected in source waters globally and have come to the public's attention. Research has shown that some microconstituents are ubiquitous in reclaimed water and, therefore, in drinking-water sources where reclaimed water is discharged. Treatment plants and soil–aquifer treatment can remove many of these compounds from wastewater, but some are recalcitrant and can be detected at very low concentrations. The human-health effects of consuming drinking water containing microconstituents is not yet known, and additional research is needed to address this issue.

2.3   Health and Regulatory Challenges

In light of the various health and regulatory challenges facing indirect potable reuse projects, Chapter 3 presents a number of recommendations to consider when planning and designing them. Such recommendations include:

- Reliable treatment processes and multiple barriers (e.g., dilution with other sources of water, setback distances, and a specified residence time underground or in surface water systems to allow for both additional treatment and time to respond if contamination occurs) should be fundamental parts of the project.
- From a regulatory perspective, the level of treatment that should be provided depends on a number of factors (e.g., the type of project; health concerns; water-quality concerns; state requirements related to indirect potable reuse; groundwater regulations, water-quality protection, and anti-degradation policies; and individual permit requirements).
- Monitoring programs must be able to verify treatment performance, detect potentially harmful contaminants, and provide the information needed to ensure that the project meets applicable regulatory requirements, to optimize treatment, and to implement pretreatment programs.

- Few states have adopted indirect potable reuse regulations. Chapter 3 presents information on the regulations that have been developed or drafted, along with approaches used by states without such regulations. The development of more state regulations or national standards of practice may advance the implementation of indirect potable reuse projects, because project implementation would be aided by knowledge of the specific regulatory requirements involved. Public acceptance of indirect potable reuse may depend, in part, on having regulations in place.

3.0   CHAPTER 4—TREATMENT TECHNOLOGY

The treatment processes needed to produce a suitable reclaimed water depend on the specific method used to augment potable water resources. The water-quality requirements for reclaimed water used to augment potable water resources are not well defined but may include limits to levels of pathogens, nutrients, trace organics, trace metals, total dissolved solids (TDS), and microconstituents. The treatment processes required also depend on the level of industrial pretreatment provided upstream of the water reclamation facility.

After conventional preliminary, primary, and secondary treatment, the processes typically used to produce reusable water include advanced treatment, natural treatment systems, disinfection, and solids management.

3.1   Advanced Treatment

Advanced treatment processes remove or further reduce constituents that remain after conventional secondary treatment.

- Biological and chemical nutrient removal. Nitrogen and phosphorus can be removed from water via biological or chemical processes. Biological nitrogen removal has two steps. In the first step, ammonia is oxidized to nitrate via nitrification. This is principally done by two groups of autotrophic bacteria. *Nitrosomonas* oxidizes ammonia to nitrite, and *Nitrobacter* oxidizes nitrite to nitrate. In the second step, nitrate replaces oxygen as an electron acceptor during biological respiration so various bacteria groups can convert nitrate into atmospheric nitrogen (i.e., denitrification). Nitrification–denitrification processes can produce effluent containing less than 3 mg/L of nitrogen.

  Coagulant addition is often the preferred method for reducing phosphorus concentrations. During biological phosphorus removal, an anaerobic zone is included in the activated-sludge reactor to select for particular strains of bacteria (e.g., *Acinetobacter*). These bacteria use short-chain volatile fatty acids (e.g., acetates), which constitute a portion of the soluble chemical oxygen demand entering the plant, and store them within the bacterial cell as polyhydroxybutyrate. The energy for this uptake is gained via hydrolysis of polyphosphates stored in these cells. During this process, hydrolyzed polyphosphates are lost from the cells into the surrounding mixed liquor as orthophosphates.

• Coagulation–flocculation and solid–liquid separation. Coagulation–flocculation and solid–liquid separation are chemical–physical processes in which a chemical (coagulant) is added to water to produce flocs of suspended particulate and to precipitate other contaminants (e.g., heavy metals, phosphorus, and organics). The chemicals used include alum, ferric chloride, lime, polymers, and various prehydrolyzed aluminum or iron salts. Each has unique properties and applications.

• Microfiltration and ultrafiltration. Microfiltration and ultrafiltration use vacuum or pressure to drive water through hollow-fiber or flat-sheet membranes, leaving pollutants behind. Microfiltration removes all particulate larger than approximately 0.1 μm. It is an effective barrier to bacteria, *Giardia*, and *Cryptosporidium*, so it can reduce disinfection requirements and, therefore, the formation of disinfection byproducts. Ultrafiltration removes all particulate larger than 0.002 to 0.1 μm; it typically is used to remove oils, colloids, and large-molecular-weight organics. It also is used as a pretreatment step for nanofiltration and reverse osmosis.

• Ion exchange. In ion exchange, electrostatically charged ions on the surface of a solid (resin) are exchanged for similarly charged ions in a solution passing across that surface. The resin can be made of natural or synthetic materials, although synthetics are more common today because of their durability. Ion exchange typically is used to remove "hardness" ions (Ca^{2+} and Mg^{2+}) from domestic water and to remove nutrients and metals from wastewater.

• Electrodialysis/electrodialysis reversal. Electrodialysis systems typically treat brackish water with low TDS concentrations; it only removes ionized compounds. Neutral dissolved organic compounds pass through them and must be removed via other means. Electrodialysis is a membrane-based process that uses electric current (a direct-current electrical field), rather than pressure, as its driving force. There are two types of electrodialysis membranes: anion membranes, which transfer negative ions, and cation membranes, which transfer positive ions. Electrodialysis systems require chemical addition to prevent scaling and fouling of membrane surfaces.

• Reverse osmosis/nanofiltration. Reverse osmosis and nanofiltration are membrane filtration processes that physically separate contaminants from water. They use pressure to drive water through the membranes. The rejection of salts (TDS), total organic carbon, and ionic species vary based on the membrane, which is classified according to the size of solutes (particles) removed. Proper pretreatment of feedwater is essential to prevent membrane fouling. Once fouling occurs, the membranes must be cleaned.

- Chemical oxidation. Chemical oxidation converts undesirable chemicals into ones that are neither harmful nor objectionable. It often is used to control both inorganics and organics. The choice of oxidizing agent [e.g., ultraviolet (UV) light, hydrogen peroxide, or ozone] may be limited by the water's intended use (to avoid residual toxic or other deleterious effects). Other considerations include treatment effectiveness, cost, ease of handling, compatibility with preceding or subsequent treatment steps, and nature of the oxidation operation.

- Advanced oxidation processes. Advanced oxidation processes are oxidation processes that involve the generation of highly reactive intermediates (e.g., radicals, especially OH radicals). Combining two oxidants increases the oxidation potential; this treatment can be more successful than using a single oxidant.

- Activated carbon adsorption. In advanced wastewater treatment, granular activated carbon adsorption primarily is intended to remove refractory and residual organics. The process works by directing water over activated carbon, which has an extremely high specific surface area for adsorbing soluble organic compounds. It also can provide some trace inorganics reduction by adsorbing inorganics chelated with organic compounds.

- Air stripping. Air stripping removes volatile organic compounds, ammonia, and carbon dioxide from water. In this process, water is pumped to the top of a tower packed with media and as it flows through the media, contaminants are volatized and exhausted. To remove ammonia, the water's pH is raised to between 10.8 and 11.5 and large quantities of air are circulated through a tower to provide air–water contact and agitation. Volatile organic compound removal relies on the tendency for moderately soluble organic compounds to leave the liquid phase and enter the vapor phase (air). Air stripping typically follows membrane treatment to remove excess carbon dioxide from water.

3.2   Natural Treatment Systems

Natural treatment systems can remove or further reduce the constituents in reclaimed water that remain after conventional secondary treatment and even after advanced treatment. The following is a brief overview of natural treatment systems.

- Constructed wetlands. Constructed wetlands are designed to mimic natural wetlands; they treat reclaimed water via emergent plants (e.g., cattails, reeds, and rushes). Constructed wetlands can reduce concentrations of organic carbon, suspended solids, nitrogen, phosphorus, metals, certain trace organics, and pathogens. There are two types of constructed wetlands: free water surface systems and subsurface-flow systems. In free water surface systems, water flows shallowly [0.3- to 0.9-m (1- to 3-ft) deep] in constructed basins or channels over a soil surface. In a subsurface-flow system, water flows underground in a trench or bed filled with porous media (typically gravel). The choice of system depends on such factors as economics, land availability, and non-treatment objectives.

- Soil–aquifer treatment. In a soil–aquifer treatment system, reclaimed water leaves the recharge basin by infiltrating the soil and moving through the subsurface to the aquifer. Various mechanisms in soil (e.g., filtration, biological degradation, physical sorption, ion exchange, and precipitation) can improve water quality. These mechanisms are effective in removing organic carbon, nitrogen, phosphorus, suspended solids, pathogens, trace metals, and trace organic compounds. Recharging groundwater with reclaimed water is an increasingly valued practice for augmenting potable water supplies, especially in arid areas.

- Riverbank filtration. In riverbank filtration (RBF), abstraction wells are installed along the bank of a river or lake, and the pumping creates a groundwater gradient from the river to the wells. When siting RBF systems, it is important to conduct a detailed hydrological and geological analysis of the proposed site. There is evidence that RBF removes pathogens, nutrients, and total organic carbon from water; decreases the potential of disinfection byproducts and membrane fouling in subsequent treatment; and has the potential for regrowth in distribution systems.

3.3   Disinfection

Wastewater typically is disinfected via chemical means (e.g., chlorination or ozonation) or photochemical means (e.g., ultraviolet irradiation).

- Chlorination. Chlorination is the most common disinfection method in the United States because of its low cost, reliability, and detectable residual. Its effectiveness is a function of the product of contact time and chlorine residual. One disadvantage of this method, however, is the potential formation of carcinogens (e.g., trihalomethanes) caused by the reaction of chlorine with organics in wastewater. In gaseous form, chlorine is extremely hazardous. Because of safety concerns, many communities now use liquid chlorine compounds (e.g., sodium hypochlorite).

- Ozone disinfection. Ozone disinfection is becoming an increasingly popular alternative to chlorination. Although ozone primarily has been used to disinfect drinking water, recent advances in ozone generation and solution technology have made it more economical for wastewater treatment. The recent awareness that ozonation byproducts can form when ozone is used in water treatment has led to a number of compounds being identified as such byproducts. Some of them could have adverse health effects.

- Ultraviolet irradiation. In recent years, UV irradiation has increasingly been used to disinfect reclaimed water; it can be effective in coliform reduction and virus inactivation. Ultraviolet irradiation emits energy in a particular wavelength range that damages organisms' genetic material. Most UV system designs consist of a reactor (closed vessel or open channel) with a number of lamps submerged in the liquid to be treated. The radiation dose is determined based on residence time in the reactor and radiation intensity. Radiation intensity diminishes with distance, so it is important that intimate contact between the liquid and lamps be maintained at all points in the reactor. The radiation intensity also is affected by the quality of the water to be disinfected.

3.4 Treatment Challenges

Many treatment-process combinations may be used to meet an indirect potable reuse project's water-quality objectives. When selecting treatment processes, it is important to follow the concept of multiple barriers. In addition, pilot-scale testing and research should be performed before the final design to ensure that water-quality objectives can be met at a reasonable cost.

4.0 CHAPTER 5—SYSTEM RELIABILITY

In terms of quantity, reclaimed water is one of most reliable sources of water available. To secure reliability, water reclamation facilities must be designed to continuously produce high-quality reclaimed water. The overall system must have multiple barriers, backup options, and flexibility.

Once a treatment system is constructed and in operation, water quality must be managed to ensure reliability. Sampling, monitoring, and permit compliance are important aspects of overall reliability. One key to making sure that system reliability is achieved is to create procedure manuals that succinctly identify permit requirements, proper operations, and prompt responses to potential upsets.

4.1 Reliable Design and Operations

- Treatment facilities. The fundamental goal of treatment system reliability is to provide reclaimed water that meets or exceeds predefined water-quality and -quantity standards. The U.S. Environmental Protection

Agency has developed wastewater-treatment reliability classes based on the receiving water's intended use and the probable adverse effect of inadequately treated discharges. The agency's most stringent class (Class I) should be considered the minimum standard for indirect potable reuse applications, unless state regulations are more restrictive. Class I reliability requires redundant facilities to prevent treatment upsets during power and equipment failures, flooding, peak loads, and maintenance shutdowns.

Reliability requirements include provisions for alarms, standby power supplies, readily available replacement equipment, multiple or standby treatment units, emergency storage or disposal and re-treatment provisions (for inadequately treated wastewater), and flexible piping and pumping facilities. A conservative design approach; properly trained, certified operators; and adequate staff are necessary to provide multiple-barrier protection and adhere to regulatory guidelines. Enough redundancy should be provided to prevent any one component from becoming absolutely vital to the protection of public health. Process piping, equipment arrangement, and unit structures must allow for O&M efficiency and ease, as well as maximum operational flexibility. Also, all aspects of plant design must allow for routine maintenance without diminishing treated water quality. Pipes or pumps that would allow bypass of critical treatment processes should not be installed. The facility should be able to operate during power failures, peak loads, equipment failure, treatment plant upsets, and maintenance shutdowns.

From a public-health standpoint, adequate, reliable disinfection is the most essential feature of the treatment process. Whether the facility uses chlorine, ozone, or UV irradiation, the disinfection process should include features to increase its reliability.

In some systems, physical separation techniques (e.g., membrane filtration) are critical in ensuring a safe water source. So, similar reliability considerations should be made: equipment redundancy, on-line monitoring of performance, and overall membrane and treatment integrity.

- Emergency storage and discharge. Reclaimed water systems should provide a reliable means of diverting water away from the potable raw-water storage and delivery system whenever the water does not meet regulatory requirements or the project's water-quality goals. To the extent possible, diversion or emergency storage must be automatically actuated via in-line sensors, which monitor critical water-quality parameters. Such monitoring would occur after each major process in the treatment train. Because reliable in-line monitoring equipment may not be available for all water-quality parameters, a manual diversion protocol based on a routine sampling program also should be established.

- Transmission and conveyance. Pipelines must be identified properly to prevent accidental damage or an unintentional cross-connection.
- Instrumentation, controls, and alarms. The choice of instrumentation depends on the reclaimed water's use. Because the potential risks of indirect potable reuse are high, instruments need to be sensitive and reliable enough to immediately detect even minor discrepancies in water quality. Provisions should be made to automatically treat, store, or dispose of reclaimed water until corrective actions have been made.
- Power supply. A standby power source should be available to energize all critical process components, including key monitoring and emergency-storage diversion or discharge facilities. The standby power should have sufficient capacity to provide services during failures of the normal power supply.
- Operations. Water treatment facilities are designed with multiple barriers to keep pathogens and harmful organic and inorganic contaminants out of drinking-water systems to the greatest extent possible. In water reclamation facilities, these barriers include secondary treatment, tertiary treatment, disinfection, soil–aquifer treatment, and surface water treatment facilities. Effective monitoring and operating procedures are other barriers.

Monitoring and operating procedures for indirect potable reuse facilities should go beyond the permit minimum. Each barrier should be described in the facility's O&M manual. The discussion should focus on the barrier's purpose, operating details, limitations, and potential for penetration.

Operations personnel should be cross-trained in each facility's division and procedures. Cross-training not only provides backup personnel, which is particularly important at smaller facilities, but also provides an awareness of other divisions' concerns. Lines of communication should be established before reclaimed water is used and regularly monitored for effectiveness thereafter.

Any number of emergencies may occur at a water reclamation facility, and rapid, effective responses depend on careful planning. Emergency response plans should include the development of a mutual aid agreement, which sets forth terms and procedures for securing personnel, equipment, and other resources from adjacent agencies. The agreement should identify each agency's emergency response coordinators and outline the various means of communication available. It is also important to establish internal lines of communication that identify whom to notify, and in what order, because all emergencies do not require the same response.

Without training, even the best plans are useless. Training should be done frequently for in-house emergencies and periodically for multi-agency emergencies.

4.2   Ensuring Water Quality

This section is intended to help the project team tailor a water-quality management plan for indirect potable reuse.

- Regulatory and performance compliance. A successfully operated indirect potable reuse system has established performance measures to ensure compliance not only with regulatory requirements but also with design, political, managerial, or public expectations. Performance criteria can be developed from design criteria; key operating characteristics; permit requirements; and political, managerial, and public performance requirements. Each treatment process should have at least one performance measure; some may be new, unfamiliar, or unique. One important aspect of performance compliance is establishing and maintaining a database, which provides the necessary background information to monitor long-term trends and helps identify developing impacts. However, performance compliance is a system-management tool; it is not intended to provide all the information needed to completely analyze a process if a violation occurs.

  Once a performance-compliance program has been established, monitoring must take place to measure compliance. Monitoring is accomplished via sampling, which must occur in two environments: in the treatment facilities and in the field.

  Also, the sampling program may be subject to more direct public observation and scrutiny than typical water or wastewater sampling programs are, so it is important to preserve a well-maintained sampling location and written procedures. Samples should be taken by trained staff who are prepared to respond directly to public inquiries during the process.

- Performance contingencies. Unless a significant violation occurs, performance excursions in water and wastewater treatment systems are primarily the concern of an operator or manager and the regulatory agency. In a potable reuse system, however, there may be considerable public concern about even minor excursions. To ensure public confidence, special consideration must be given to how violations are defined and corrected in a potable reuse system.

  The political, management and public standards are the most critical because they are developed to alleviate the community's specific concerns. A violation may be an immediate confirmation to the public that its concerns were justified.

  Violations are always a high priority, but they are even more so in a potable reuse system, so it is important that the notification process and corrective actions include the public as well as the regulatory agency.

A procedures manual may be an indirect potable reuse system's most important public relations document. It can help build public confidence if written in a user-friendly way, so information can be provided quickly before misinformation begins to circulate.

- Regulatory Agency Participation. Indirect potable reuse is relatively new, so there is a lack of long-term, scientifically documented data on which to base regulations. If design and discharge criteria are not available, regulations may have to be project specific. So, including regulators as part of the project team may prove beneficial.

5.0 CHAPTER 6—ADDRESSING PUBLIC PERCEPTIONS

Although indirect potable reuse projects can provide many benefits to both communities and water utilities, implementing them requires special attention to public perceptions. Project success depends on a utility's ability to address people's concerns. Effective public outreach can reduce the likelihood of infrastructure underinvestment, facility abandonment, damage to the utility's reputation and relationships, and negative branding. It also can increase the likelihood that water-reuse benefits are considered fairly, that reuse proposals are supported adequately, and that reuse projects are chosen when they offer the best value to the community.

Community outreach is intended to develop relationships and gain community support when solving problems. First, the problem must be identified (e.g., the need for a new water supply or the need to ensure water reliability for the region or community). The problem must be identified and understood before discussing a solution, because the community may not support a proposed project if they do not fully embrace the problem first. Solving the problem is the reason for proposing an indirect reuse project, the reason for the community to accept perceived risks, and the meaningful context of the outreach effort.

Once the community recognizes the problem, the utility should propose multiple solutions. The public needs to see that the utility is committed to solving the problem, not weeded to a specific solution. The goal is for the community to invest appropriately to solve the problem. If an indirect potable reuse project is proposed, its success will likely be influenced by the public's perception of risk, trust in the utility, and any related "branding." Indirect potable water reuse may be a compelling solution because it provides communities with a new, high-quality, drought-resistant water supply that can be stored in existing reservoirs or aquifers.

In addition, water utilities should understand the relationship between utilities and politicians. Politicians are highly visible and understandably want to avoid risk, protect their reputations, and maintain their communities' support.

Politicians should feel that the water utility understands this position and will help uphold the community's trust and confidence.

Utilities also should develop relationships with several key groups of individuals: potential opponents, elected officials, city employees, active community members, business leaders, ethnic and social group leaders, local regulators, the media, trusted medical or technical community leaders, and other individuals who are looked to for community leadership. These relationships should be pursued diligently—even when it is difficult to maintain communications. The outreach effort is not about convincing people to change their minds, but to develop the relationships and trust needed for the community to make the best decision about ensuring future water reliability.

The water utility should become the public's primary source of information about water. Although the public typically understands that many common activities have associated risks, they also need to understand how the risks related to indirect potable water reuse compare and how the utility is managing such risks. Ideally, a community that believes its water utility manages risk well will trust that utility. Community members may not be exposed to everything the utility does, but they see how the utility's representatives behave at public meetings or other events and develop their trust accordingly. Water utility communications should convey knowledge, diligence, and care. Other actions (e.g., implementing aggressive testing protocols, meeting and exceeding regulatory requirements, providing multiple barriers to contamination, and maintaining system redundancy) also can inspire trust in a utility. This trust makes the utility the source of water quality, not the physical source of the drinking water (a critical objective when the source is wastewater).

The perception that water quality depends on the source has been glamorized routinely in advertisements. When indirect potable reuse projects are being discussed, a utility may need to emphasize that a water's quality can be independent of its source. Advanced treatment technologies and monitoring methods can be used to build public confidence in water quality, regardless of source.

Throughout the communication process, the water utility must discuss all options openly and candidly. Conflicting views should be embraced as potential sources of new ideas and better community investments. Opposing views should be addressed early, when utilities and engineers still can make changes to their proposals. Collaborating with project opponents may draw attention to water issues, which can end up improving the overall project. Conducting an outreach program with no intention of listening to community members and considering changes is a quick way to incite conflict and erode trust in the utility. Sometimes opposition can be quelled via dialogue (e.g., the opposing parties can resolve a misconception about the project). In other cases, opponents may not change their views but may respect the utility's position if they understand the constraints under which it makes decisions. Besides garnering public support, managing conflict can help a utility gain the trust of politicians and policy makers.

Whether a community trusts its water utility can be heavily influenced by "branding". In essence, a *brand* is a set of widely held, specific notions about something (e.g., a commercial product, an organization, or a situation); it may be positive or negative and can have a lasting effect.

Case studies—both successes and failures—illustrate how real utilities worked to communicate with the communities they serve. Utilities in Arizona, California, Florida, and Virginia all pursued indirect potable reuse projects but had varying degrees of success in earning their communities' trust; conveying key project objectives to the public; inspiring confidence in water quality; uniting individuals with opposing viewpoints; and supporting politicians so they, in turn, could support the project.

Based on these utilities' experiences, a timeline for implementing indirect potable reuse projects has been developed. This timeline suggests that utilities should lay the initial foundation for project success 7 to 8 years before construction is scheduled to begin. This foundation should consist of building rapport, developing a solid water-quality reputation, and establishing a habit of effective dialogue with the community.

Then, specific relationships should be developed, and support for general ideas should be garnered. After approximately 4 years of foundation building, project design can begin. If utilities effectively establish themselves as capable of managing risk, negotiating conflict, and maintaining trust, this can lead to the construction of valuable indirect potable reuse projects.

Chapter 2

Planning Indirect Potable Reuse

1.0   INTRODUCTION

Water has always been reused via the global hydrologic cycle. As water scarcity and demands on available renewable supplies of water increase, the intimacy of water reuse increases (i.e., the local hydrologic cycle is forcibly shrunk, and the proximity of actual reuse gets closer to the point where the water previously was considered a waste). Most people in developed countries, including the United States, use water resources that include some water that has been used at least once before.

Planning and managing water resources in basin-scale watersheds and within utility service boundaries are complex tasks that involve a variety of technical, institutional, and public health and policy issues. The complexity of planning increases as the demands on water increase. While planned water reuse was born out of a desire to have an alternative to discharge, it has evolved into a practice necessary to ensure that the water resources portfolio is robust. Intensive water use typically results in the indirect potable reuse of water—whether planned or unplanned. There probably are no remaining "raw" water sources, including deep aquifers, serving the drinking water needs of large populations that are completely free of previously used water.

Following are definitions of some terms typically used when discussing water recycling or water reuse:

- Reclaimed water. Also called "recycled water," *reclaimed water* is water that has been used at least once (wastewater) and has been adequately treated for another use.
- Indirect potable reuse. *Indirect potable reuse* is the process of deliberately introducing some amount of reclaimed water to a natural water system (e.g., a river, reservoir, or aquifer) that is a source of drinking water supplies. The purpose is to augment existing drinking water supplies.
- Direct potable reuse. *Direct potable reuse* is the process of using some amount of reclaimed water as a source of drinking water supplies without introducing it to a natural water system first.
- Unplanned reuse. *Unplanned reuse* is the incidental use of reclaimed water for any purpose (i.e., without thoroughly considering the applicable and interrelated technical, economic, regulatory, and public policy issues beforehand).
- Planned reuse. *Planned reuse* is using reclaimed water for predefined beneficial purposes after having thoroughly considered the applicable and interrelated technical, economic, regulatory, and public policy issues.

Direct potable reuse currently is not practiced in the United States. Unplanned indirect potable reuse is a common situation in watershed basins or utility service areas. Planned indirect potable reuse is becoming increasingly common in the

United States in recognition of the regulatory gap that exists between the Clean Water Act and the Safe Drinking Water Act. Planned indirect potable reuse acknowledges the physical link between discharges of wastewater and production of water supply and incorporates publicly acceptable barriers between the two.

This chapter presents some of the major issues that a water resource planner should consider when planning an indirect potable water reuse project. Their applicability depends on whether the plans are at the basinwide or utility level.

2.0 INTEGRATED WATER RESOURCES PLANNING

In most developed countries, including the United States, it is increasingly necessary to provide water-related services as part of an integrated water resources plan because there are very few instances in which water, wastewater, reclaimed water, and stormwater utilities do not affect one another, the environment, or recreational water uses. Theoretically, considering the effects and relationships among water-related activities in a watershed or a utility service area:

- Reduces the potential for negative environmental effects,
- Encourages a more complete understanding of drought effects and development of adequate responses to drought or other interruptions in water services,
- Reduces the total cost of service to rate payers,
- Takes into account concerns related to climate changes,
- Addresses water-quality issues more holistically, and
- Maximizes the use of local water resources.

This chapter addresses indirect potable reuse planning from an integrated water resources perspective.

In general, reclaimed water is reliable in terms of quality, quantity, and continuous supply, so it is an important resource to consider for augmenting potable water resources. Because it is continuously available, it should be considered equal in value to groundwater and surface water from a volumetric standpoint. In its 2003 report, *Water Reuse for Florida: Strategies for Effective Use of Reclaimed Water*, the Florida Reuse Coordinating Committee and the Water Conservation Initiative's Water Reuse Work Group ranked indirect potable reuse (including groundwater recharged with reclaimed water) among the highest desirable reuse activities because they can use 100% of the available reclaimed water to augment surface water and groundwater supplies. This ranking reflects the potential value of indirect potable reuse in integrated water resources management. Presumably, indirect potable reuse would be considered because traditional drinking water sources lack sufficient quantity, quality, or reliability, or because of restrictions related to reclaimed water management, such as effluent discharge restrictions or seasonal inability to make use of reclaimed water for nonpotable demands.

As with most endeavors that require science and engineering to resolve social problems, there are two questions that need to be answered:

- "Can it be done?" (i.e., is it technically feasible and cost-effective?), and
- "Should it be done?" (i.e., is it publicly acceptable?).

The answers will determine whether indirect potable reuse should be considered as one approach to addressing the concerns about the total water supply. Other approaches to consider—in addition to indirect potable reuse—include better water delivery and use efficiency (supply- and demand-side conservation), nonpotable reuse, exchanges of reclaimed water for other sources of drinking water supply, and acquisition and development of new sources of supply.

In summary, the decision to implement indirect potable reuse (e.g., augmenting surface water or groundwater supplies for potable use) should be made within the framework of an understanding of the total watershed or utility service area, the technical feasibility, cost, and public acceptance of all the alternatives being evaluated to resolve water supply problems.

3.0   MEANS OF INTRODUCING RECLAIMED WATER TO POTABLE WATER SUPPLIES

There are several means of introducing reclaimed water into water supplies. A primary consideration is the "intimacy" between the reclaimed water and the drinking water supply. While there are numerous examples of successful indirect potable reuse projects, there are also many examples of failed ones. Failures are largely attributed to negative branding by detractors, who perpetuated negative perceptions and fear of a source of water once considered a "waste". U.S. successes typically introduced a natural system between the water reclamation facility and the drinking water treatment facility. For example, West Palm Beach, Florida, conveys highly treated reclaimed water through a wetlands that recharges the regional aquifer used for drinking water purposes.

Reclaimed water may be introduced to potable water supplies via natural systems by one or both of the following methods:

- *Groundwater recharge*, in which reclaimed water is added to a groundwater aquifer by direct injection or by surface spreading via recharge basins or alluvial infiltration; and
- *Surface water augmentation*, in which reclaimed water is added directly to surface water supplies (e.g., reservoir, lake, stream, or river).

3.1   Groundwater Recharge

In many parts of the United States, sustained groundwater withdrawals have resulted in *groundwater overdraft*, a condition in which an aquifer's natural recharge rate cannot keep pace with the pumping rate. The direct disadvantages of groundwater overdraft include rising pumping costs (because of the need to drill deeper wells) and, eventually, a depleted aquifer. The secondary disadvantages are site specific; they have included

- Land subsidence, fissures, and associated damage to surface infrastructure and property;
- Movement of poorer quality water into the aquifer (either laterally from an ocean and/or brackish waterbody, or vertically because of wells or natural conditions that allow vertical movement); and
- Declines in surface water flows that depend on groundwater. Recharging aquifers with reclaimed water can help restore aquifer volumes and minimize the secondary effects of overdraft.

Another trend is that groundwater aquifers are being used to store water (i.e., an underground reservoir). Aquifer storage has certain advantages over surface storage (Metcalf and Eddy, Inc., 1991). Subsurface storage typically is not prone to evaporative losses, except when the water table is very shallow or tapped by *phreatophytes*—deep-rooted plants that obtain water from the groundwater supply.

The aquifer itself can function as a water conveyance system, providing lateral movement without pipelines. Delivery from groundwater storage can be increased or provided at a new location by drilling more wells. Depending on site-specific conditions, this storage and delivery method may have significantly lower capital and operations and maintenance (O&M) costs than a traditional aboveground storage and distribution facility.

Water-quality degradation can be a concern in either surface or aquifer storage systems. Surface water sources typically are more readily cleansed of contamination than aquifers, but aquifers are less susceptible to rapid and extensive contamination, unless they are directly connected to surface water sources (e.g., surficial gravel unconfined aquifers).

Water can be introduced to groundwater aquifers via injection wells, recharge basins, or alluvial infiltration. There are three types of injection wells: direct-injection wells, aquifer storage and recovery (ASR) wells, and vadose zone recharge wells. *Direct-injection wells* introduce reclaimed water directly to the aquifer. *Aquifer storage and recovery wells* not only deliver reclaimed water directly to the aquifer but also can recover water from the aquifer. *Vadose zone recharge wells* inject reclaimed water into the unsaturated zone between the ground surface and the water table of the saturated aquifer.

Recharge basins are shallow basins that receive reclaimed water and allow it to percolate down to the water table.

Alluvial infiltration involves releasing water into a stream channel and allowing it to percolate to the water table. The concept is similar to recharge basins and considered a variation of them elsewhere in this publication. Where appropriate and if regulations allow, stream channels can be modified with partial dykes to detain flow, thereby encouraging more recharge in a shorter reach of the channel.

Figure 2.1 shows cross sections of groundwater-recharge methods. The choice of recharge method depends on the following factors:

- Land availability, topography, and cost;
- Climatic factors (e.g., evaporation rate, temperature, and wind);
- Regulatory and permitting constraints;
- Location of potable water wells relative to injection wells;
- Subsurface geology and sustainable recharge rates;
- The aquifer's hydrogeology;
- Degree of mixing and the groundwater's subsurface flow velocities;
- Soluble materials and contaminants (e.g., nitrates) in the subsurface strata;
- The reclaimed water's quantity, quality, and flow characteristics;

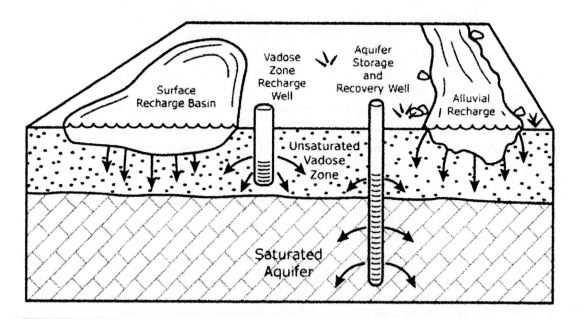

FIGURE 2.1   Cross-section view of various groundwater-recharge methods.

- Chemical characteristics of both recharge water and aquifer water;
- Cost of facilities;
- Environmental impacts;
- Compatibility with regional and community quality-of-life goals;
- Ability to recover the stored water; and
- Public policy and input.

When siting and sizing injection wells, a designer should consider land availability and current land use. Pipelines and wells typically can be accommodated because pipelines can be buried, while well-head facilities typically need a fraction of an acre per facility. Also, well houses may be made to aesthetically blend into various landscape and architectural settings, and well heads can be constructed in subsurface vaults.

Using direct-injection wells circumvents the concerns of intervening impermeable layers (strata) with leachable salts or minerals. Also, injection wells do not contribute to evaporative losses because they lack an exposed free-water surface. Properly designed direct-injection wells send water directly and immediately into the aquifer. There is no time lag (as is the case when the water must percolate to the water table) and no opportunity for obstruction or quality degradation in transit. However, once the reclaimed water is injected, there is no opportunity for additional treatment.

Aquifer storage and recovery wells have gained popularity because of improvements in associated technology and operating regimes. They allow users to better manage the diurnal and seasonal fluctuations in reclaimed-water supply and demand.

Vadose zone recharge wells can be used to "punch through" impermeable geologic layers to introduce the recharge water to more transmissive geology. They also can be used to take advantage of the subsurface geology's natural treatment capacity before introducing the water to the aquifer.

Designers also should consider topographic and current land-use constraints when siting and sizing recharge basins. The potential for onsite soil treatment may reduce pretreatment requirements before recharge. Depending on the volume of water to be recharged and the percolation rate possible through the underlying strata, the recharge basins may need to cover tens or hundreds of acres.

Sometimes, surface spreading basins are the preferred recharge alternative because of the desire to create riparian habitat and wildlife viewing opportunities. For others, the exposed water surface is a concern because it could attract waterfowl and propagate midges or mosquitoes. Moreover, recharge basins could have significant water losses, depending on their exposed surface area and the atmosphere's evaporative demand. Atmospheric depositions could change water quality over time. Large basins with poor circulation and "dead spots" may encourage the growth of nuisance algae.

For recharge basins to be effective, a hydrologic connection to the target aquifer must exist. Also, subsurface materials beneath the basin must be relatively permeable and free of impermeable layers that might impede the downward movement of recharge water. If the target aquifer is overlain by less-permeable material, it may not be possible to recharge the deeper water-bearing zones via basins. If the strata between the basin and aquifer—or within the aquifer itself—contain water-soluble salts and minerals, they might leach and degrade water quality. Surface soils in areas with a sustained history of human activity are more likely to contain agricultural and industrial contaminants (e.g., pesticides, nitrate, and volatile organic compounds). Recharge could "push" or carry these contaminants from the soil into the aquifer, so when siting recharge facilities, previous land use should be diligently assessed.

In addition to water-quality considerations, groundwater recharge projects can have other environmental impacts. For example, the reclaimed water intended for groundwater recharge may have historically augmented a surface waterbody, and removing it from that waterbody may harm the related ecosystem because of its dependence on the reclaimed water. Wetlands-recharge, riverbank filtration, or surface basin recharge projects with riparian habitat can have positive environmental impacts, but maintaining their flows and volumes becomes important not just for water-supply management purposes but also for environmental, aesthetic, and recreational purposes. Also, creating habitats in such projects opens the possibility of endangered and protected species inhabiting the project, which can significantly influence how the facility must be managed. (For more information, see the "Regulations" section of this chapter.)

The quantity and timing of the water supply are important factors when sizing recharge basins and injection wells. If the groundwater-recharge project is part of an integrated water management scheme, it is possible that the availability of reclaimed water for recharge will vary depending on the other demands for it. Also, some communities have large seasonal population changes or inflow and infiltration, which can cause the quantity of reclaimed water being produced to fluctuate. Large fluctuations in reclaimed-water quantity and timing typically result in recharge facilities that are not used at their full-functioning capacity. Low utilization reduces the efficiency of recharge facilities and may adversely affect their construction and operation economics. Increasingly, recharge facilities are designed to accommodate multiple sources of water (e.g., unused surface water allocations, urban stormwater, and reclaimed water from industrial sources or neighboring utilities) to maximize the facilities' utility and economics.

The volume of reclaimed water to be introduced to an aquifer is a fundamental consideration in indirect potable reuse projects. This amount relative to the aquifer's "native" groundwater determines the percentage of reclaimed water in the raw water supply. The project team must understand the aquifer's hydrogeology so the reclaimed-water fraction can be estimated and used for planning and communicating with regulators and the public.

The quality of the water supply is an important factor when selecting a recharge method. Recharge water directly injected into a potable supply aquifer typically must meet or exceed all federal and state drinking water standards. Lower-quality water typically can be used in recharge basins because of the natural filtration and purification processes that occur as the water percolates down to the aquifer. The common assumption that the passage of source water through the soil to the aquifer and through the aquifer to the withdrawal point provides no treatment is overly conservative when applied to most chemicals and microorganisms (NRC, 1994). Based on current knowledge, however, municipal wastewater that will be reclaimed for use in recharging aquifers should receive at least secondary treatment (biologically treated and clarified) if being spread in a surface basin. It should receive secondary and tertiary treatment (filtered and disinfected) if being injected directly into the aquifer. Some states' statutes and codes provide specific treatment or water-quality requirements for indirect potable reuse projects.

Scientists are just beginning to understand the fate, transport, and potential health effects of a wide variety of microconstituents in treated wastewater (some of which originate in pharmaceutical compounds and personal care products). Until the risks of these microconstituents are better understood, their presence at least must be acknowledged and considered in any plans to augment potable water resources. This is why flexibility and space are recommended for designs of indirect potable reuse projects—particularly to accommodate additional and advanced treatment at both the reclamation and reuse points.

These natural filtration processes also reduce concerns about the chemical compatibility of recharge water and native groundwater. Such compatibility is critical to the operation of a well-injection system. If the waters' chemical compositions are unstable, precipitates may form, thereby reducing the recharge facility's capacity. Well-injection systems also are sensitive to plugged wells, well screens, and the aquifer by bacterial growth. Properly pretreated injection water and periodic well maintenance typically will address problems associated with chemical incompatibility or bacterial growth.

The costs associated with either recharge basins or injection wells are typically an important factor in deciding whether recharge is a viable storage option and which recharge methods to use. The ability to use existing production wells is an important cost consideration, as are the costs of land acquisition, facility construction, and annual operations and maintenance. Cost estimates and economic analyses for aquifer storage of reclaimed water should be developed using the same procedures followed to evaluate any other source of supply. Groundwater storage systems (using either recharge basins or injection wells) have proven to be economically feasible in many locations for decades.

Even with good hydrogeologic data and accurate water-quality data, it is difficult to predict with certainty how recharge processes will behave. Constructing a pilot recharge and recovery facility (with instrumented monitoring wells) is strongly recommended to determine the feasibility of recharge (Viessman and Hammer, 1993). Problems arising in the relatively inexpensive pilot facility can be identified, diagnosed, and, if possible, mitigated in the full-scale design.

3.2   Existing Groundwater Recharge Facilities

To date, potable-supply aquifers have been successfully recharged with reclaimed water via both well injection and recharge basins. Table 2.1 is a list of recharge projects using reclaimed water that have been operating for a number of years.

A more recent project that augments groundwater supplies with reclaimed water is the Water Campus in Scottsdale, Arizona, which uses direct injection and ASR wells after advanced treatment and blending with other water resources.

3.3   Surface Water Augmentation

Using reclaimed water to augment surface water (Figure 2.2) is a practice that is gaining favor among water purveyors. By recognizing and accounting for the deliberate introduction of reclaimed water to water supplies, agencies in water-short communities can supplement their surface water supplies.

The practice of augmenting surface water with reclaimed water must be done carefully and with thorough evaluation of several important considerations.

3.3.1   Locations of the Delivery and Intake Points

The locations of the pipes that deliver reclaimed water to a surface water and that withdraw it for drinking water purposes (i.e., the delivery point and the intake point) are important. Depending on the distance, significant attenuation of water quality can occur between these two points. *Attenuation* is the biological degradation of residual wastes, adsorption, sedimentation, and exchange of residual contaminants; however, not all contaminants may be effectively attenuated. Intervening streams (both in and out), groundwater effects, and other natural changes in the watercourse can also change the raw water's character.

The locations of the delivery and intake points also can affect a stream's flow pattern. In reservoirs, the short-circuiting of reclaimed water to the intake point should be minimized.

In some cases, rather than adding reclaimed water to a reservoir, it may be more appropriate to introduce it to the surface water below the reservoir and use an equivalent amount of surface water generated above the reservoir for drink-

TABLE 2.1   Pioneering recharge projects.

Project	Location	Agency	Method	Startup	Capacity
Montebello Forebay	Los Angeles, California	County Sanitation Districts of Los Angeles County	Groundwater recharge—surface spreading basins	1962	50 000 AFY (45 mgd)
Water Factory 21	Fountain Valley, California	Orange County Sanitation District and Orange County Water District, who operate the Ground Water Replenishment Project (GWRS).	Groundwater recharge—direct injection	1976	9400 AFY (8.5 mgd)
Hueco Bolson	El Paso, Texas	City of El Paso	Groundwater recharge—direct injection	1985	4000 AFY (3.5 mgd)

Research Project	Location	Agency	Method	Duration	Capacity
San Diego water repurification	San Diego, California	City of San Diego	Surface water augmentation	1964–present	1 mgd
Denver potable water demonstration project	Denver, Colorado	City of Denver	Direct reuse	1985–1992	0.1 mgd
Tampa water resource recovery project	Tampa, Florida	City of Tampa	Surface water augmentation	1983–present	N/A

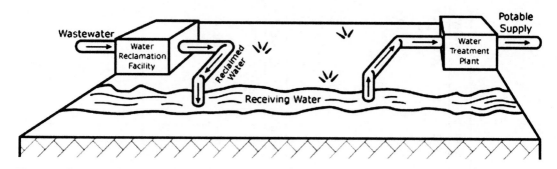

FIGURE 2.2   Schematic of surface water augmentation.

ing water purposes. Such augmentation may allow for more mixing with other raw water supplies that enter the surface water system below the reservoir. It also encourages natural attenuation between the introduction point and the next downstream use.

3.3.2   Volume of Water To Be Reused

The volume of reclaimed water to be introduced to the water supply is a fundamental consideration for indirect potable reuse projects. The percentage of reclaimed water in the raw water supply depends on the relative amounts of reclaimed and "native" water in a receiving stream or reservoir. It is important to understand the variations in hydrologic flow patterns over time to determine both the average and extreme conditions (e.g., droughts). The project team then can calculate the fraction of reclaimed water that will be in the water supply and use this for planning and communicating with regulators. To date, no acceptable ranges of raw-to-reclaimed-water ratios associated with indirect potable reuse have been published. The acceptable ratio will be project specific and established based on feasibility, cost, need, and public acceptance of all the alternatives being considered.

When augmenting reservoirs, a water resource planner should consider and assess the low water level and whether reclaimed water will mix thoroughly with raw water. Nutrient load and concentration management are important because algae growth can contribute to clogging of water treatment plant intakes and filters and to possible taste and odor concerns. Depending on the specific system, the seasonal drawdown of reservoirs can create an extreme condition. If so, structural or operational modifications may be required so reclaimed water will mix fully with raw water.

3.3.3   Character of Watershed

The project team should evaluate the water quality of the watershed in which indirect potable reuse will be practiced. This evaluation should consider all water uses, including drinking. It is important to consider the cumulative effect of all potential sources of water (e.g., other introductions of treated wastewater or nonpoint sources of pollution) on total water quality.

In addition, water-quality monitoring must be performed to adequately characterize background conditions. Microbiological, organic, and inorganic parameters should be measured at enough locations and frequencies to establish the expected water quality if reclaimed water is introduced to the water supply. Also, monitoring programs should be designed to reflect the seasonal and human-induced hydrologic variations that may occur. The background monitoring period should be defined in terms of months, years, flow rate, injection rate, travel time, etc. to ensure that the reclaimed water's residence and travel times have been considered.

4.0   REGULATIONS

Legal and regulatory issues at both the state and federal levels present many challenges to the implementation of potable water reuse. Several regulatory considerations (e.g., regulatory and permitting issues, water rights issues, and case law) must be factored into the planning of an indirect potable reuse project.

4.1   Federal Regulations

At the federal level, a number of key laws and regulations affect the use of reclaimed water, although none specifically address potable reuse. These include the following:

- The Safe Drinking Water Act. The Safe Drinking Water Act is designed to protect the nation's drinking water by establishing standards for drinking water quality. These standards may directly or indirectly constrain indirect potable reuse.
- U.S. EPA's Long-term 2 Enhanced Surface Water Treatment Rule. The Long-term 2 Enhanced Surface Water Treatment Rule (LT2ESWTR) requires that utilities dependent on surface water sources provide filtration and adequate disinfection to inactivate viruses and protozoan cysts. It also addresses systems that rely on groundwater that is directly influenced by surface water. So, depending on the quality of the reclaimed water used to augment surface waters, the LT2ESWTR may require more stringent water treatment because of the possible presence of microorganisms.

- U.S. EPA's Stage 2 Disinfectants and Disinfection Byproducts Rule. The Stage 2 Disinfectants and Disinfection Byproducts Rule stringently regulates trihalomethane compounds and other potentially carcinogenic organic compounds in drinking water supplies. The concentrations of such compounds partly depend on the presence and concentration of natural organic matter in the source water and may constrain indirect potable reuse projects.
- U.S. EPA's Groundwater Rule. The Groundwater Rule uses a targeted, risk-based strategy to address the fecal contamination risks of untreated groundwater via surveys of public water supplies, water-quality monitoring, disinfection or other mitigation if necessary, and compliance monitoring. This rule might constrain the ability to use reclaimed water for aquifer recharge or cause an existing variance to be revoked.
- The Clean Water Act. The Clean Water Act directs the U.S. EPA to help states implement groundwater, surface water, and wetland protection strategies. In some cases, discharge standards promulgated under the National Pollutant Discharge Elimination System (NPDES) permitting program may be so stringent as to favor indirect potable reuse rather than discharging treated wastewater to a receiving waterbody.
- The National Environmental Policy Act. The National Environmental Policy Act requires an environmental evaluation of every project that requires federal action or relies on federal funding and might significantly affect the environment. Water projects typically fall under the auspices of this act.
- The Endangered Species Act. The Endangered Species Act seeks to conserve rare and endangered species. Federal agencies must carry out programs for creatures on the list of threatened and endangered species and ensure that the projects they authorize, fund, or undertake are not likely to jeopardize the continued existence of such species. Water projects frequently fall under the auspices of this act. Care should be taken during the planning and permitting processes to ensure that reuse-related augmentation does not become a sustained and rigid obligation to maintain water flow associated with critical habitat for threatened and endangered species. Otherwise, the water's usefulness for drinking may become limited. It may be necessary, however, to maintain stream flows with some portion of the reclaimed water supply because the habitat became dependent on years of prior releases of reclaimed water, or because of water exchange arrangements.

Other important federal laws and regulations to consider include the Fish and Wildlife Coordination Act, Coastal Zone Management Act, Wild and Scenic Rivers Act, and National Historic Preservation Act.

4.2   State Regulations

States typically have parallel regulations to most of these federal rules and regulations. In some cases, state standards may be more restrictive than the federal ones (Getches et al., 1991). While a good summary of each state's water reuse regulations is provided in U.S. EPA's *Guidelines for Water Reuse* (2004), state regulations are expected to be updated regularly as states gain experience with and confidence in water reuse.

Every state with water reuse rules developed them within the framework of the existing rules related to surface water, groundwater, drinking water, and wastewater quality and quantity. So, they vary significantly from state to state. State rules also can be difficult to find and understand because they often do not use the phrase "indirect potable reuse" (or something similar) and frequently are embedded or implied in regulations not specifically related to water reuse or wastewater treatment.

It is essential that project planners consult with state and local departments of natural resources, environmental control, health, etc. very early in a water reuse project to gain a clear understanding of the rules and regulations that will apply to the project. Some states (e.g., Arizona, California, and Florida) have rules related to planned indirect potable reuse, although they may be called something else.

In Arizona, for example, groundwater recharge is governed by the aquifer protection permit program under the Arizona Department of Environmental Quality (regardless of the source of recharge water). It also is governed by water-storage and groundwater-storage facility permits under the Arizona Department of Water Resources. Reuse permits are issued for specific uses associated with appropriate and defined qualities of reclaimed water, but groundwater recharge and stream augmentation are not included in this list. Arizona codes and statutes do not specifically address augmentation of potable water resources, but the regulatory framework allows it.

Several states have passed legislation concerning the nondegradation of groundwater via well injection, surface basin recharge, and some irrigation practices. These laws vary from state to state. Also, several states have regulations related to nondegradation of surface water that are more stringent than the NPDES permitting requirements.

4.3   Water Rights

A *water right* typically is defined as the right to use water, not the right of ownership. Water rights issues can constrain reclamation and reuse projects by imposing restrictions and requirements on the reuse and return of water (U.S. EPA, 2004). Because of the complexities of water rights issues and differences in state water rights laws, project planners should seek expert guidance during the early stages of a reclamation and reuse project. Some broad descriptions of the complexities are provided below.

State legislatures and courts established water rights doctrines in response to the unique hydrologic conditions prevalent in their state (Rice and White, 1987). Federal doctrines are important in the body of water rights laws, but to a large extent, state doctrines form the fundamental core of water law.

The key federal doctrines of water law that affect water apportionment (and so could constrain the use of reclaimed water) are as follows:

- The *equitable apportionment doctrine*, implemented via interstate compacts or U.S. Supreme Court decree, applies to states whose boundaries are crossed by a stream. It assigns each state a share of such interstate waters, creating an upper limit on the amount of water in them that the state may allocate internally. Each state may allocate its share in accordance with its own legal doctrines, but water reuse is not allowed to violate the apportionment.
- The *reserved right doctrine* provides that when the United States reserves land from the public domain (e.g., creates a national forest or park), it also reserves enough water to ensure that the land's primary new purpose is not defeated (Weinberg and Allan, 1990). Such reserved rights carry the same priority date as that of the land reservation, meaning that the right to use water is applicable from the date at which the land was set aside and not from the date on which water was first used. If reusing reclaimed water (rather than returning it to a stream) could injure a reserved right, then reuse is not allowed.

Under state water rights laws, the key factors affecting doctrines that control water use are whether the water is surface water or groundwater, and the natural abundance of water in the state.

4.3.1 Surface Water Doctrines

4.3.1.1 Appropriation Doctrine

For surface waters in states with scarce water resources, an *appropriation doctrine* was developed that allowed streams to be used beneficially by anyone, but under a strict system of priorities.

When the priorities are established based on the water-diversion date, this is called the *prior appropriation doctrine*, which is based on the concept that "first in time is first in right." Under this system, the person with the first priority is entitled to take and use as much water as his or her quantified right allows (the owner's full right) to the extent that physical water availability allows. If water still remains, then the person with the second priority may divert water up to his or her full right, etc. In this system, a *right* constitutes the beneficial use of water but requires that whatever volume of water that is not fully consumed (e.g., municipal wastewater or agricultural runoff) be returned to the stream. The prior appropriation doctrine is a significant impediment to water reuse projects. To implement reuse projects under this doctrine,

- Exchange arrangements must be developed with other stream users,
- Reuse systems must be operated only when there is no need to return municipal wastewater to the stream to fulfill prior rights, or
- Only special categories of water (e.g., transbasin imports or nontributary groundwater) that do not have to be returned to the stream must be used.

Transbasin-imported water is water diverted from one basin and introduced to another. Because this water would not be available if not for the diverter's efforts, the diverter typically is entitled to use and reuse it completely ("to extinction"). *Nontributary groundwater* is water pumped from an aquifer. Because this water would not have contributed to the stream flow except for the pumper's efforts, the pumper typically is entitled to use and reuse it to extinction.

4.3.1.2 Riparian Doctrine

For surface water in states with more abundant water resources, a *riparian doctrine* was developed that allowed landowners next to a stream (*riparian owners*) to make reasonable use of the water. Under this doctrine, a riparian owner can divert and use water if the use is reasonable and made on riparian land. In each riparian-doctrine state, different laws and court rulings have been implemented to help define *reasonable use* and *riparian land*. The ease or difficulty of developing a water reuse project under riparian doctrine rules depends largely on how a state defines these terms.

4.3.1.3 Mixed or Hybrid Doctrines

In certain states with varying degrees of water-resource abundance, a mixed system has developed that includes aspects of both appropriation and riparian doctrines. These states are described as *mixed doctrine* or *hybrid doctrine* states. The rules governing the development of water reuse projects in such states will vary from state to state and, in some cases, by location within the state.

4.3.2 Pueblo Rights

In some states (e.g., California and New Mexico), communities or cities still may be subject to Pueblo water rights, which were established under Spanish law in former Mexican or Spanish territories. Pueblo rights typically are superior to other water rights (Rowe and Abdel-Magid, 1995) and provide significant latitude in the development of potable reuse projects.

4.3.3 Groundwater Doctrines

Several groundwater-use doctrines have been developed. The *absolute privilege doctrine* (also called the *absolute ownership doctrine*) essentially decrees that a landowner owns all water beneath the surface of his or her land and is entitled to take it regardless of the consequences. The *reasonable use doctrine* allows landowners to pump all the water they can, as long as it is used on the overlying land for reasonable and beneficial purposes. The *correlative rights doctrine* is an extension of the reasonable use doctrine; it allows scarce groundwater to be shared in proportion to the extent of overlying land ownership. The *prior appropriation doctrine* is similar to the surface water doctrine of the same name. In some cases, it puts groundwater withdrawals in the same priority system as surface water diversions. Several states have implemented permit systems and require that a permit be obtained specifying location and amount of groundwater withdrawal.

The key aspect of these groundwater doctrines for potable reuse projects is the ability to establish ownership of the water recharged to an aquifer for storage. If there is no reasonable guarantee that the recharged water will not be used by others, then using aquifers for storage is not an attractive option. In many states, specific legislation is being considered to address this concern—not only for potable reuse purposes, but for aquifer storage and conjunctive use projects in general.

5.0   ECONOMICS

Many issues affect the financial feasibility of indirect potable reuse projects. Such issues include water-quality goals or standards, the costs of alternatives, the project financing options, and the method for reimbursing O&M costs. One particularly important issue is that an investment already has been made in developing the original water resource (which is now reclaimed water).

A good resource for evaluating the benefits and costs of water reuse is the WateReuse Foundation's *An Economic Framework for Evaluating the Benefits and Costs of Water Reuse* (2006). It includes a CD-ROM with the tools necessary to store and evaluate project-specific economic factors.

5.1   Water-Quality Goals

In indirect potable reuse projects, water-quality goals for both the reclaimed water and treated drinking water should be established; these will affect the project's economics. At a minimum, the water must comply with all applicable regulatory requirements. The National Primary Drinking Water Standards consider raw water quality a fundamental element of drinking water systems.

Producing drinking water that poses no threat to the consumer's health depends on continuous protection. Because of human frailties associated with protection, the purest source should be chosen. Polluted sources should not be used unless other sources are economically unavailable, and then only when the personnel, equipment, and operating procedures can be depended on to purify and otherwise continuously protect the drinking water supply. A multi-barrier approach is appropriate to minimize opportunities for introducing hazardous levels of contaminants to humans. This approach may increase costs, so opportunities for expanded benefits associated with other community or basinwide goals (e.g., creating urban wetlands for wildlife viewing or establishing urban-recreation lakes) should be sought. For more information on water-quality criteria that apply to planned indirect potable reuse systems, see Chapter 3.

5.2 Treatment Requirements

When considering an indirect potable reuse project, readers should investigate all the related treatment or water-quality requirements. The treatment requirements for reclaimed water used in indirect potable water reuse projects are driven by state and local statutes and rules. There are no specific federal treatment or water-quality standards for indirect potable water reuse. If there are no specific state rules, however, one can rely on the Clean Water Act and the Safe Drinking Water Act as guidance for attaining the treatment and water-quality requirements necessary to protect public health and the environment.

Indirect potable reuse projects typically must address two treatment facilities: the one producing reclaimed water and the one producing drinking water. Producing reclaimed water often involves more processes than those conventionally used at wastewater treatment facilities to comply with NPDES requirements. Such processes include soil–aquifer treatment, wetlands polishing, membrane filtration, advanced oxidation processes, and granular activated carbon filters.

Conventional water treatment plants typically include chemical addition, flocculation, sedimentation, filtration, and disinfection processes. These are intended to remove suspended solids and kill pathogens found in the water source. Unfortunately, conventional water treatment does not remove many constituents of concern that may be found in municipal wastewater (e.g., carcinogens, pharmaceutically active compounds, and certain microbiological organisms). So, more treatment may be required to remove microbial and chemical constituents from the blended water supply if their concentrations limit indirect potable reuse and they persist despite the water reclamation process, the dilution and attenuation processes that naturally occur in waterbodies and aquifers, and the conventional water treatment process. More treatment also may be required to address specific concerns raised by the public (this may be more than the regulations require).

5.3   Costs of Facilities

A primary consideration of indirect potable reuse programs is its cost and how it compares to those of other water-supply alternatives. The overall project cost should include both direct and indirect costs (e.g., those related to administrative, legal, and public-acceptance efforts). It is important to consider all associated costs, so a fair evaluation may be made.

Another consideration is which entity will ultimately bear the project's costs. This is relatively simple if one entity controls both wastewater and water utilities, but if different entities control them, cost sharing may become complex. For example, if it is determined that additional treatment is necessary for one or more water quality constituents, the water purveyor may decide to accept financial responsibility for the incremental cost of treatment at a wastewater treatment plant because it is less expensive to treat at the wastewater treatment plant than to treat at the water treatment plant. The proper division of the costs in such circumstances is difficult.

6.0   PUBLIC PARTICIPATION AND ACCEPTANCE

The ultimate success of any water reuse program is determined by its level of public acceptance. Social psychologists have long known that it is difficult for humans to reconcile that something once contaminated (e.g., wastewater) could ever be cleaned enough to be used again. This phenomenon is called *contagion psychology* (i.e., once contaminated, always contaminated). So, to gain public acceptance, it is important that the reclaimed water "lose its identity" before being consumed (Metcalf and Eddy, Inc., 1991; U.S. EPA, 2004). This loss is most successfully achieved by installing multiple barriers:

- Adequately treating the reclaimed water,
- Introducing it to a natural system,
- Diluting it with water already in the natural system (to acceptable levels), and
- Withdrawing the mixture from the natural system and consistently treating it to meet drinking water standards.

Gaining public acceptance requires a well-conceived program of public involvement and education. An essential component is informing people why indirect potable water reuse needs to be included in the overall water supply plan. This discussion should address the

- Need for more water;
- Availability, reliability, and costs of alternative supplies;
- Concept of conservation as it relates to reclamation;

- Information on successful potable and nonpotable reuse projects elsewhere;
- Economic, environmental, and other social benefits;
- Available health risk information; and
- Design criteria and treatment technologies used in water reclamation plants.

Demonstration pilot programs work well in communicating to the public the effectiveness and reliability of water reclamation plants. For a helpful tool in building the public trust needed to ensure that indirect potable water reuse projects receive fair consideration, see www.watereuse.org/Foundation/resproject/WaterSupplyReplenishmt/index.htm (Water Supply Replenishment).

Water reuse has been practiced for as long as upstream users have been returning flows to streams from which downstream users divert water. The act of consciously reusing water and doing so closer to its original use must be considered carefully. Public acceptance of this practice depends largely on how the water is reused.

Reclaimed water typically is well accepted for a variety of nonpotable uses. Reusing water for industrial purposes (e.g., processing and cooling) typically is embraced by the public and touted by industrial public relations departments as an environmentally friendly practice. Reusing water to irrigate landscapes is accepted and, in many cases, encouraged by the public. Reusing water to irrigate agriculture also is well accepted—particularly for livestock forage crops or food crops that must be peeled or husked before human consumption (e.g., oranges and sweet corn). Reusing it to irrigate other food crops (e.g., tomatoes and apples) is acceptable when there are adequate safeguards against bacterial or viral contamination, and extensive educational programs.

Indirect potable reuse (e.g., introducing reclaimed water to a stream, reservoir, or aquifer used for potable supply) typically has been well accepted by the public when accompanied by extensive informational and educational programs to describe the purity standards and public health safeguards involved. Educated communities have come to recognize the value of water resources and the roles of water reclamation and potable reuse in environmentally and economically sound resource management and sustainable development.

Direct potable reuse, however, currently does not have significant support from the public, regulators, water purveyors, water treatment professionals, or water-marketing agencies and companies (e.g., those involved in bottled water, point of use, or point of entry). While direct and near-direct potable reuse has been evaluated in Denver, Colorado; Tampa, Florida; and San Diego, California, full-scale implementation has not yet occurred in the United States. On the other hand, Windhoek, Namibia, has been practicing direct potable reuse since 1968; depending on water-source conditions, reclaimed water makes up to 25% of its drinking water. Singapore's NEWater project also practices direct potable reuse;

reclaimed water has made up 1% of local drinking water supply since 2003, and plans are to increase this to 2.5% by 2012. Research and these successful projects show that an informed public can advocate direct potable reuse (Lohman, 1987). For more information on gaining public acceptance, see Chapter 6.

7.0 REFERENCES

Getches, D. H., MacDonnell, L. J.; Rice, T. A. (1991) *Controlling Water Use: The Unfinished Business of Water-Quality Protection*; Natural Resources Law Center, University of Colorado School of Law: Boulder, Colorado.

Lohman, L. C. (1987) Potable Wastewater Reuse Can Win Public Support. *Proceedings of Water Reuse Symposium IV*; Denver, Colorado, Aug. 2–7; American Water Works Association Research Foundation.

Metcalf and Eddy, Inc. (1991) *Wastewater Engineering: Treatment, Disposal, Reuse*; 3rd ed.; McGraw-Hill: New York.

National Research Council (NRC) (1994) *Ground Water Recharge Using Waters of Impaired Quality*; National Academy Press: Washington, D.C.

Reuse Coordinating Committee and Reuse Work Group of the Water Conservation Initiative (2003), *Water Reuse for Florida: Strategies for Effective Use of Reclaimed Water*. Florida Department of Environmental Protection: Tallahassee, Florida.

Rice, L.; White, M. (1987) *Engineering Aspects of Water Law*; Wiley and Sons: New York.

Rowe, D. R.; Abdel-Magid, I. M. (1995) *Handbook of Wastewater Reclamation and Reuse*; Lewis Publishers: New York.

U.S. Environmental Protection Agency (2004) *Guidelines for Water Reuse*; EPA/625/R-04/108; Technology Transfer: U.S. Environmental Protection Agency: Cincinnati, Ohio.

Viessman, W.; Hammer, M. J. (1993) *Water Supply and Pollution Control*; 5th ed.; Harper Collins College Publishers: New York.

WateReuse Association (2007) *Water Supply Replenishment: Creating a New Source of High Quality Water*; www.watereuse.org/Foundation/resproject/WaterSupplyReplenishmt/index.htm (accessed March 2008).

Weinberg, E.; Allan, R. F. (1990) Federal Reserved Water Rights. In *Water Rights of the Fifty States and Territories*; American Water Works Association: Denver, Colorado.

8.0 SUGGESTED READINGS

Jain, R. K.; Urban, L. V.; Stacey, G. S.; Balbach, H. E. (1993) *Environmental Assessment*; McGraw-Hill: New York.

Lohman, L. C.; Milliken, J. G. (1985) *Informational/Educational Approaches to Public Attitudes on Potable Reuse of Wastewater*; Denver Research Institute, University of Denver: Denver, Colorado.

Pyne, D. (1995) *Groundwater Recharge and Wells: A Guide to Aquifer Storage Recovery*; Lewis Publishers: Boca Raton, Florida.

U.S. Environmental Protection Agency (1980) *Wastewater in Receiving Waters at Water Supply Abstraction Points*, Washington, D.C.

WateReuse Foundation (2006) *An Economic Framework for Evaluating the Benefits and Costs of Water Reuse*; 03-006-02; WateReuse Foundation: Alexandria, Virginia.

Chapter 3

Health and Regulatory Considerations

1.0   INTRODUCTION

This chapter presents an overview of health and regulatory issues that should be considered when planning and implementing an indirect potable reuse project. These issues are typically more complex than those encountered in direct non-potable reuse projects primarily because the perceived and actual risks of using reclaimed water for indirect potable projects are higher than those for non-potable reuse applications. Indirect potable reuse projects, which deliberately supplement public drinking water with reclaimed water, have an inherently higher potential health risk than nonpotable reuse projects (e.g., irrigation) in which ingestion typically would be infrequent and incidental. So, nonpotable reuse applications primarily are regulated for microbiological parameters, while indirect potable reuse also must address the potential health implications of long-term ("chronic") ingestion of recycled water. In addition, indirect potable reuse projects may face challenges related to water quality or environmental concerns associated with the release and co-mingling of reclaimed water with surface water or groundwater.

The traditional maxim for selecting drinking water supplies has been to use the highest quality source available (U.S. EPA, 1976). However, based on need or cost, many cities take water from rivers or other surface waters that receive wastewater discharges and use conventional water treatment (e.g., filtration and disinfection) to eliminate the pathogens responsible for waterborne diseases. Other communities use groundwater influenced by land disposal of wastewater or septage. Even though federal regulations for protecting groundwater from fecal contamination were not issued until 2006 (U.S. EPA, 2006c), U.S. water utilities long have provided water that meets current drinking water regulations.

The practice of diverting raw water supplies downstream of wastewater discharges is often called *incidental or unplanned potable reuse*. It occurs in thousands of U.S. communities, including Philadelphia, Cincinnati, and New Orleans, which draw water from the Delaware, Ohio, and Mississippi Rivers, respectively.

Since the 1950s, secondary treated wastewater has been discharged into Nevada's Las Vegas wash, which currently provides 2% of the inflow into Lake Mead, Las Vegas Valley's primary drinking water source.

Even with the prevalence of unplanned indirect potable reuse, some experts believe that water originating from raw municipal wastewater is inherently more risky—even if the reclaimed water meets drinking water standards (NRC, 1994, 1998). Their concerns include

- Disinfecting reclaimed water may create different and often unidentified disinfection byproducts than those found in conventional water supplies;
- Only a small percentage of the organic compounds in drinking water have been identified, and the health effects of only a few of these have been determined;
- The health effects of mixtures of two or more of the hundreds of compounds in any reclaimed water used for potable purposes are not easily characterized; and
- The whole process relies on technology and management.

While these issues are not insurmountable, they illustrate the need to acknowledge that complex health and regulatory issues—related to public health, water quality, and environmental protection—must be evaluated and accommodated in indirect potable reuse projects. Many of these concerns are common to drinking water (i.e., not unique to reclaimed water). This chapter provides guidance on how to negotiate the increasingly complex regulatory environment for indirect potable reuse projects.

2.0 HEALTH ISSUES
2.1 Microorganisms

Literally hundreds of types of pathogens are excreted in the fecal material of infected hosts and enter municipal wastewater (Rao and Melnick, 1986; Straub et al., 1993). Diseases caused by waterborne microorganisms range from mild gastroenteritis to severe illnesses, such as infectious hepatitis, cholera, typhoid fever, and meningitis (Feachem et al., 1983; Mead et al., 1999). For infection to occur, the infectious agent must be present in the community; it must survive all of the wastewater treatment processes to which it is exposed; an individual must directly or indirectly come into contact with reclaimed water; and the water must contain enough of the agent to cause infection at the time of contact. The percent of infections that result in disease (symptoms) varies from pathogen to pathogen (Soller et al., 2004).

The principal infectious agents in wastewater can be classified into three broad groups: bacteria, parasites (protozoa and helminths), and viruses. Table 3.1

TABLE 3.1   Infectious agents that could be present in raw wastewater (Hurst et al., 1989; Sagik et al., 1978).

Pathogen	Disease
Bacteria	
Campylobacter jejuni	Gastroenteritis
Escherichia coli (enteropathogenic)	Gastroenteritis
Legionella pneumophila	Legionnaire's disease
Leptospira (spp.)	Leptospirosis
Salmonella typhi	Typhoid fever
Salmonella (2 400 serotypes)	Salmonellosis
Shigella (4 spp.)	Shigellosis (dysentery)
Vibrio cholerae	Cholera
Yersinia enterocolitica	Yersiniosis
Protozoa	
Balantidium coli	Balantisiasis (dysentery)
Cryptosporidium parvum	Cryptosporidiosis, diarrhea, fever
Entamoeba histolytica	Amebiasis (amebic dysentery)
Giardia lamblia	Giardiasis
Helminths	
Ancylostoma duodenale (hookworm)	Ancylostomiasis
Ascaris lumbricoides (roundworm)	Ascariasis
Echinococcus granulosis (tapeworm)	Hydatidosis
Enterobius vermicularis (pinworm)	Enterobiasis
Necator americanus (roundworm)	Necatoriasis
Schistosoma (spp.)	Schistosomiasis
Strongyloides stercoralis (threadworm)	Strongyloidiasis
Taenia (spp.) (tapeworm)	Taeniasis, cysticercosis
Trichuris trichiura (whipworm)	Trichuriasis
Viruses	
Adenovirus (51 types)	Respiratory disease, eye infections
Astrovirus (5 types)	Gastroenteritis
Calicivirus (2 types)	Gastroenteritis
Coronavirus	Gastroenteritis
Enteroviruses (72 types) (polio, echo, coxsackie, new enteroviruses)	Gastroenteritis, heart anomalies, meningitis, others
Hepatitis A virus	Infectious hepatitis
Norwalk agent	Diarrhea, vomiting, fever
Parvovirus (3 types)	Gastroenteritis
Reovirus (3 types)	Not clearly established
Rotavirus (4 types)	Gastroenteritis

lists many of the infectious agents that may be present in raw municipal wastewater and their related diseases.

The occurrence and concentration of pathogens in raw wastewater depend on a number of factors, including the wastewater's sources, the contributing population's general health, the existence of disease carriers in the population, and the ability of infectious agents to survive outside their hosts in various environmental conditions. Moreover, it typically is impossible to predict with any degree of confidence what the general characteristics of a particular wastewater will be with respect to infectious agents.

Left untreated, raw wastewater may be a significant public health risk. Current treatment technologies allow for significant pathogen reduction. Indirect potable reuse projects that use soil–aquifer treatment (SAT) can remove more pathogens, including viruses and parasites. *Soil–aquifer treatment* is the natural physical, chemical, and biological processes that further treat the water as it infiltrates down through the soil to groundwater.

However, the ability to routinely measure specific pathogens in treated wastewater is limited by the availability of reliable and sensitive analytical methods. Surrogate parameters such as coliforms still must be used to characterize desired treatment levels. The advent of newer, molecular-based analyses—particularly polymerase chain reaction (PCR)—may be much more sensitive, but more work is needed to link results and organism viability before these methods can be used for compliance monitoring or risk assessment.

2.1.1　Bacterial Pathogens

Bacteria are microscopic organisms ranging from approximately 0.2 to 10 μm long. They are ubiquitous in nature and have a wide variety of nutritional requirements. Many types of harmless bacteria colonize the human intestinal tract and are routinely shed in feces. Pathogenic bacteria are present in the feces of infected individuals, so municipal wastewater can contain wide varieties and concentrations of bacteria.

2.1.2　Protozoan Parasites

In general, protozoan parasite cysts are larger than bacteria, ranging in size from 2 to more than 60 μm. While parasitic cysts are present in the feces of infected individuals, they also can be excreted by healthy carriers (those with no discernable clinical disease). Like viruses, cysts do not reproduce in the environment but can survive there for extended periods of time (e.g., more than 6 years in soil under ideal conditions) (Bryan, 1974). Parasites and bacteria typically can be removed effectively after percolation through a short distance of the soil mantle.

Florida requires *Giardia* and *Cryptosporidium* to be monitored in reclaimed water (FDEP, 1999). Results for facilities that provide secondary treatment, filtration, and disinfection are presented in Table 3.2.

Studies in Florida and elsewhere routinely found *Giardia* cysts and *Cryptosporidium* oocysts in reclaimed water that had undergone filtration and high-level disinfection (Walker-Coleman et al., 2002; Rose and Carnahan, 1992; Sheikh and Cooper, 1998; Rose et al., 2001; Rose and Quintero-Betancourt, 2002; York et al., 2002) and was deemed suitable for public access uses (U.S. EPA, 2004). A number of more detailed studies that considered the viability and infectivity of the cysts and oocysts suggested that *Giardia* probably was inactivated by

TABLE 3.2   Summary of Florida pathogen monitoring data (total cysts or oocysts per 100 L)(Walker-Coleman, 2002, © 2002 WateReuse Association).

Statistic	Giardia	Cryptosporidium
Number of observations	69	68
Percent with detectable concentrations	58%	22%
25th percentile (#/100 L)	ND*	ND*
50th percentile (#/100 L)	4	ND
75th percentile (#/100 L)	76	ND
90th percentile (#/100 L)	333	2.3
Maximum (#/100 L)	3096	282

*ND = nondetect.

chlorine, but 15 to 40% of detected *Cryptosporidium* oocysts may survive (Keller, 2002; Garcia et al., 2002; Gennaccaro et al., 2003 ; Sheikh et al., 1999; Quintero-Betancourt et al., 2003). Other studies evaluating UV irradiation and the electron beam as alternatives to chlorine disinfection found that both parasites were easily inactivated, and both *Giardia* cysts and *Cryptosporidium* oocysts were completely inactivated by a UV dose less than 10 mJ/cm^2 (Mofidi et al., 2002; Slifko, 2001).

2.1.3   Viral Pathogens

Viruses are obligate intracellular parasites that can only multiply within a host cell, where they are assembled as complex macromolecules via the cell's biochemical system. Viruses occur in various shapes, range in size from 0.01 to 0.3 μm in cross section, and are composed of a nucleic acid core surrounded by an outer coat of protein (Bitton, 1980). *Bacteriophages* are viruses that infect bacteria (they have not been found to infect humans).

Viruses typically are not excreted for prolonged periods by healthy individuals, and their occurrence in municipal wastewater fluctuates widely. Hosts typically shed viruses at a rate of 1000 to 100,000 infective units/g of feces; virus shedding has been reported to be as high as 10^9/g of feces (Cooper, 1975; Feachem et al., 1983). Enteric virus density in undisinfected secondary effluent from municipal wastewater treatment plants can range from 3.5 to 75 plaque-forming units (PFU)/1000 L (NRC, 1998). Virus densities in wastewater are typically seasonal; the highest concentrations frequently are found in summer and early autumn. Viruses typically are more resistant to environmental stresses than are many bacteria, although some viruses persist for only a short time in municipal wastewater. The Sanitation Districts of Los Angeles County's historical virus monitoring of filtered disinfected effluent has shown an average 3-log inactivation, based on secondary effluent concentrations and assuming a theoretical detection limit of 1 PFU/1000 L (300 gallons) (Yanko, 1993). The Sanitation Districts now have evaluated more than 1200 samples of filtered disinfected effluent, of which 1199 were reported to be below detectable virus limits. [For a numerical method of interpreting negative virus results from highly treated water, see Crohn and Yates (1997).]

Viruses can survive underground for extended periods and have been isolated by a number of investigators examining a variety of recharge operations after various migration distances. Soil type and composition, pH, moisture content, and virus strain all interact to affect the adsorptive capacity and virus dieoff rate in soil (Goyal and Gerba, 1979; Powelson et al., 1993). Viruses have been documented to migrate up to 46 m (150 ft) vertically through soil and more than 366 m (1 200 ft) horizontally in groundwater at land-disposal sites with permeable soils (Gerba and Goyal, 1985). Soil aquifer treatment of filtered disinfected effluent, using bacteriophage as a tracer, has shown that a 7-log reduction in bacteria could be expected to occur as the water traveled within approximately 30 m (100 ft) through the subsurface (Fox et al., 2001).

2.2 Waterborne Illness

The infectious doses of selected pathogens (i.e., numbers of organisms necessary to initiate infection) are presented in Table 3.3. Information is also available on the functional forms, distributions used to describe dose–response parameters, and the dose–response parameters for specific pathogens used in microbial risk assessments, as presented in Table 3.4.

Approximately 85 to 90% of all non-foodborne infections in the United States, including those believed to be from waterborne sources, are thought to be caused by viral pathogens (e.g., enteric viruses) (Mead et al., 1999). The relative

TABLE 3.3 Infectious doses of selected pathogens[a] (Feachem et al., 1981; 1983).

Organism	Infectious dose* CFU, PFU, or Cysts/Oocysts per 100 L
Escherichia coli (enteropathogenic)	10^6–10^{10}
Clostridium perfringens	1–10^{10}
Salmonella typhi	10–10^7
Vibrio cholerae	10^3–0^7
Shigella flexneri 2A	180
Entamoeba histolytica	20
Shigella dysentariae	10
Giardia lamblia	<10
Cryptosporidium	1–10
Ascaris lumbricoides	1–10
Viruses	1–10

*Some of the data for bacteria are ID_{50}—the dose that infects 50% of the people given that dose. Persons given lower doses also could become infected. CFU = coliform-forming unit. PFU = plaque-forming unit.

TABLE 3.4   Summary of pathogen dose–response relations (Soller et al., 2007).

Pathogen	Dose–response form and endpoint	Parameter distribution	Value(s)	Value(s)	References
Rotavirus	Hypergeometric (Infection)	Point estimates	$\alpha = 0.167$	$\beta = 0.191$	Teunis and Havelaar, 2000
Cryptosporidium spp.	Exponential (Infection)	Uniform	$r_{lower} = 0.04$	$r_{upper} = 0.16$	U.S. EPA, 2006a,b,c
Giardia spp.	Exponential (Infection)	Point estimate	$r = 0.0199$		Rose et al., 1991; Teunis et al., 1996
E. coli O157:H7	Hypergeometric (Infection)	Point estimates	$\alpha = 0.08$	$\beta = 1.44$	Teunis et al., 2004
Salmonella spp.	Gompertz log (Illness)	Uniform	$\alpha_{lower} = 29$	$\alpha_{upper} = 50$	Coleman and Marks, 1998, 2000; Coleman et al., 2004; Oscar, 2004
		Point estimate	$\beta = 2.148$		

importance of viral pathogens in waterborne diseases is supported by data from the World Health Organization (WHO, 1999) and research conducted during the last 20 years on exposure to waterborne pathogens via recreational activities (Cabelli, 1983; Fankhauser et al., 1998; Levine and Stephenson, 1990; Palmateer et al., 1991; Sobsey et al., 1995). Norwalk-like viruses have been reported to account for 23 million illnesses each year in the United States, of which 60% are estimated to be non-foodborne. Rotavirus accounts for 3.9 million illnesses each year in the United States, of which 99% are non-foodborne (Mead et al., 1999).

In comparison, the protozoon *Giardia lamblia* has been reported to cause 2 million illness cases per year, of which 90% are non-foodborne, and *Cryptosporidium parvum* causes 300,000 illnesses each year, of which 90% are non-foodborne (Mead et al., 1999). Many of the important bacterial illnesses in the United States are typically foodborne. Of the 5.2 million annual bacterial illness cases in the United States, *Salmonella* (*Salmonella spp.* and nontyphoidal), *Shigella*, and *Campylobacter spp.* have been reported to collectively account for about 4.3 million (approximately 95, 20, and 80%, respectively, of these infections were foodborne) (Mead et al., 1999).

Groundwater is a source of potable water supply for more than 40% of the U.S. population. Approximately 72% of the U.S. public water supply systems that use groundwater do not disinfect (Yates, 1994), but this may change as systems comply with the U.S. Environmental Protection Agency's groundwater disinfection rule (U.S. EPA, 2006a). The use of contaminated, untreated, or inadequately treated groundwater has been the cause of approximately 50% of waterborne diseases in the United States since 1920 (Craun, 1986a, 1986b, 1991; Herwaldt et al., 1992). Most of the outbreaks were caused by pathogens.

The 1993 outbreak of cryptosporidiosis in Milwaukee, Wisconsin, during which an estimated 403,000 people became ill and 4400 were hospitalized, was the largest documented waterborne disease outbreak in the United States since recordkeeping began in 1920. The outbreak was associated with filtered and chlorinated Lake Michigan water. Deteriorating raw water quality and decreased effectiveness of the coagulation–filtration process resulted in inadequate removal of *Cryptosporidium* oocysts (Kramer et al., 1996).

2.3 Chemical Constituents

All water supplies could be contaminated by three types of chemical substances—inorganics, organics, and radionuclides—that can be affected by natural physical and chemical processes (e.g., precipitation, filtration, and adsorption). These physical and chemical processes typically are acceptable means of purifying water to eliminate potential human health effects (except for mineral contamination, radon emissions, etc.) and make it potable. However, anthropogenic activities contribute innumerable sources or potential sources of contamination to natural water supplies that could increase human health risks if consumed. So, a number of water protection practices and regulations have been adopted to limit this risk.

Municipal wastewater's possible constituents may be a concern when reclaimed wastewater is designated as a source of potable water. Part of this concern has to do with the components of reclaimed water that have known health-related properties and traditionally are monitored and regulated in drinking water. There are no national standards for indirect potable reuse, so designers and regulators typically apply existing wastewater and drinking water regulations for quality, treatment, and reliability to indirect potable reuse applications. Drinking water regulations are addressing growing numbers of contaminants, so planners and designers of indirect potable reuse projects will need to prepare accordingly.

Health concerns arise with indirect potable reuse because of the unknown nature of the mixtures of materials in raw wastewater that survive or are created during treatment. These mixtures can result from typical household and industrial wastes. However, many concerns associated with organic matter in reclaimed water are analogous to those associated with natural organic matter in surface waters or groundwaters used as drinking water sources. Although a number of comprehensive studies have been designed to address the concern about potential risks of unknown and unidentified chemicals in reclaimed water (Gruener, 1979; Nellor et al., 1984), there is currently no definitive measure of risk or safety for this issue.

Organic and inorganic chemicals in reclaimed water may be dissolved or associated with particulate. Consuming them theoretically could cause acute or chronic public health effects, but their actual effects are undetermined and may be no greater than the risks related to other water supplies. Therefore, source control, treatment processes, water quality monitoring, and treatment reliability testing are relied on to limit their effects on potable water.

2.3.1 Inorganics

In general, the health hazards associated with inorganics (e.g., trace metals and nonmetals) are well established. The U.S. Environmental Protection Agency has developed maximum contaminant levels (MCLs) in drinking water for 12 metals: antimony, arsenic, barium, beryllium, cadmium, chromium, copper, lead, mercury, nickel, selenium, and thallium. The agency regulates more metals (e.g., bismuth, cobalt, silver, tin, and vanadium) via industrial wastewater pretreatment programs. The limits are in the microgram-per-liter (μg/L) to low milligrams-per-liter (mg/L) range. Metals of aesthetic interest (e.g., calcium hardness and iron) and of health concern (e.g., sodium) also are frequently monitored.

Inorganic nonmetallic constituents include anions typically present at harmless levels in natural waters that also may be of aesthetic interest or have some potential health significance at elevated concentrations (e.g., alkalinity, boron, chloride, fluoride, and sulfate). Some anions of health significance can occur at trace levels in treatment or disinfection processes (e.g., bromate, chlorite, cyanide, nitrate, and nitrite). The U.S. Environmental Protection Agency has established MCLs in the mg/L range for most of these constituents.

Trace metals enter water via natural sources (e.g., mineral formations and deposition during precipitation) or human-related activities [e.g., corrosion (the effect of aggressive water on copper plumbing) and industrial contamination]. Human-related activities are the primary concern in reclaimed water treatment. Although small amounts of a few trace metals are essential nutrients, they have toxic effects at concentrations much higher than those typically found in drinking water. Limiting reclaimed water concentrations of trace metals to drinking water MCLs is typically considered a protective procedure.

In water reuse, the nonmetals ammonia, nitrite, and nitrate are a health issue because nitrate concentrations significantly greater than 10 mg/L as nitrogen, the federal MCL, have caused a disease called methomoglobinemia in infants or debilitated individuals (Bouchard et al., 1992). Nitrogenous compounds are an inherent part of domestic wastewater, and the nitrogen cycle is a prominent part of the biochemistry of reclamation. Most forms of nitrogen can readily convert to nitrate ions either in the treatment process or in soils, so nitrogen management is a significant issue in indirect potable reuse projects.

Other nonmetallic constituents (e.g., fluoride, chloride, sulfate, bicarbonate, or boron) are essentially soluble and can result from either natural (e.g., mineralization) or industrial sources. Fluoride may be added to some drinking waters to maintain approximately 1 mg/L concentrations for dental-health purposes.

However, the federal MCL is 4 mg/L to avoid adverse health effects. Reclaimed water could be a source of fluoride contamination in indirect potable reuse if wastewater contributions of fluoride approach this concentration.

Residential use typically adds about 300 mg/L of dissolved inorganic solids to water, although the amount ranges from approximately 150 mg/L to more than 500 mg/L (Metcalf and Eddy, 2002). Most of the increase is attributable to anions and dissolved metals (e.g., sodium and potassium). In some parts of the country, chloride enters wastewater via automatic water softeners in homes or businesses when sodium chloride or potassium chloride is used to regenerate the softener resins. While the concentration of dissolved solids in reclaimed water may not be a health concern, it may make complying with water quality standards more challenging. Therefore, testing for total dissolved solids (TDS) may be a frequent monitoring requirement in addition to chemical-specific testing.

The inorganic disinfection byproducts bromate and chlorite are a primary concern in drinking water when ozone or chlorine dioxide, respectively, is used to disinfect water or wastewater. Otherwise, these anions are of little significance.

Asbestos is the common name of several mineral forms that may exist as fibers or particulates in water supplies. Primary asbestos sources are surface water contamination and asbestos–cement pipes used in drinking water distribution systems. Although U.S. EPA has established an MCL for asbestos in drinking water, the connection between waterborne asbestos consumption and disease is questionable. Therefore, asbestos has been of little interest in reclaimed water projects and is monitored infrequently, if at all.

2.3.2   Radionuclides

Radionuclides are inorganic constituents that emit alpha, beta, and gamma radiation that can cause genetic damage to specific tissues, which can lead to organ failure or tumor formation. Sources of radiochemicals in water supplies may include mineralization, aerial fallout, precipitation, and industrial or clinical wastes. Some radionuclides may be in reclaimed wastewater as a result of the discharge of regulated or unregulated isotopes from industrial or clinical sources. Therefore, radioisotope monitoring in reclaimed water is not uncommon. Although drinking water's contribution to the total body burden of radiation dosage traditionally has been considered to be low, U.S. EPA has developed MCLs for total or gross alpha emitters, gross beta emitters, and specific isotopes (Ra^{226}, Ra^{228}, and uranium). These MCLs are expressed as radiation doses in picocuries per liter (pCi/L).

2.3.3   Organics

The trace synthetic organic chemical content of potable water supplies has caused concern as a potential source of chronic human health effects, especially carcinogenic, mutagenic, or teratogenic responses to long-term exposure to low concentrations (μg/L or less). These concerns came into sharp focus during the 1970s when a large number of studies attempted to identify trace organics in water supplies via surveys of organics (e.g., pesticides and chlorinated solvents) in U.S. drinking water supplies and determine their toxic effects in animals or other test organisms. Such studies continue to proliferate. The U.S. Environmental Protection Agency has established drinking water MCLs for more than 20 volatile (solvent) organic compounds and approximately 50 pesticides, herbicides, and other industrially produced organics. The agency also has established MCLs for disinfectants and disinfection byproducts [e.g., trihalomethanes (THMs) and haloacetic acids]. Most of these regulated compounds have theoretical or known human health significance, but the problem of human health assessment of organics in water supplies is of much higher magnitude than monitoring for approximately 70 regulated compounds.

While the number of organics identified in water supplies is between 1000 and 2 000 compounds (Donaldson, 1977), detecting the number of organics that could be present may be beyond modern analytical capabilities. It has been reported that less than 10 to 20% of the total amount of organic matter in a given water sample has been—or can be—identified (Ding et al., 1996). In addition, only about 50% of the disinfection byproducts generated during water treatment have been identified (Singer, 1994). Therefore, the origins and health effects of this unidentified matter are often subjects of speculation and study, as are concerns with biochemical interactions (e.g., synergism between mixtures of organics). The toxicological risk, or safety, of these mixtures may never be known precisely. The continuous creation and use of new synthetic organic chemicals precludes comprehensive testing for all potentially toxic compounds and creates an ever-present element of uncertainty for all reclaimed water projects involving indirect potable reuse.

In the meantime, traditional methods of measuring organic matter [e.g., biochemical oxygen demand (BOD), chemical oxygen demand, total organic carbon (TOC), dissolved organic carbon, or total organic halogens (TOX)] are used. Although these constituents have no direct health significance, they are used as surrogates for monitoring the efficacy of existing processes or evaluating new organics-removal methods. Some states use TOC as a surrogate for controlling microconstituents (as discussed later in this chapter). Such analyses are part of a monitoring framework that includes analyses of specific compounds.

For reclaimed water intended for indirect potable reuse, a great deal of concern exists about the number of chemicals that could enter the reclamation system, the disinfection byproducts [e.g., THMs or N-nitrosodimethylamine (NDMA)] produced when municipal wastewater is disinfected with chlorine or chloramines (Mitch et al., 2003); the amount of unidentified organic matter; and the potential for chronic health effects if these materials contaminate drinking water supplies. While these concerns may exist for any water supply, reclaimed water typically merits more concern because of its ostensible connection with sources of pollution.

2.3.4 Microconstituents

As a result of recent analytical advances and reconnaissance monitoring studies of surface water and drinking water using these new methods, the public has become aware that endocrine-disrupting chemicals, pharmaceuticals, and personal care products have been detected in source waters worldwide. These chemicals (e.g., trace organic compounds, pharmaceuticals, personal care products, hormonally active substances, microbes, prions, and nanomaterials) are collectively called *microconstituents* because they typically are found in tiny concentrations (μg/L or less) in water.

Research has shown that some microconstituents are ubiquitous in municipal wastewater effluents and subsequently occur in drinking water sources where wastewater is discharged (Table 3.5) (Daughton and Ternes, 1999; Halling-Sørensen et al., 1998; Koplin et al., 2002; Richardson and Bowron, 1985; Stan and Heberer, 1997; Ternes, 1998).

Wastewater treatment and soil–aquifer treatment (SAT) can remove many of these compounds, but some are recalcitrant and can be detected at very low concentrations (Fox et al., 2001; Fox et al., 2006; Dickenson, 2006; Drewes et al., 2001; Drewes, 2006; Snyder, 2006). Of particular interest are compounds that can affect the endocrine system, which are called *endocrine disruptors*. A chemical's ability to bind to the estrogen receptor, either *in vivo* or *in vitro*, has been used as a definition of *estrogenicity* (Gharravi et al., 2006). Of concern is whether exposure to chemicals with steroid-like activity can disrupt normal endocrine function, leading to altered reproductive capacity, infertility, endometriosis, and cancers of the breast, uterus, and prostate. Soil aquifer treatment efficiently removed estrogenicity, as measured via *in vitro* and *in vivo* assays (Fox et al., 2006). Endocrine-disrupting compounds (EDCs) are expected to be nonpolar and biodegradable, and their transport in the subsurface is limited (Heberer et al., 2002). Additionally, as analytical methods are modified to permit the detection of ultra-trace levels of contaminants [e.g., nanograms per liter (ηg/L) or less], more compounds will be found. However, the ability to detect a compound does not necessarily translate to health concerns.

The predominant issues currently associated with microconstituents in surface waters focus on the effects on aquatic life and wildlife, particularly effects on the endocrine system (e.g., feminization of fish). However, these effects can be the result of other factors or may be natural variations in populations.

Epidemiological information on the human health effects of consuming drinking water containing microconstituents is not available, so their presence or presumed presence in reclaimed water used for indirect potable use is a concern often raised by opponents. This is a difficult challenge for project planners and designers

TABLE 3.5   Personal care products, pharmaceuticals, and endocrine disruptors detected in drinking water sources and raw and finished drinking water (prepared by Intertox, Inc., Seattle, Washington, 2006).

Drinking water sources (groundwater or surface water)		Raw drinking water	Finished drinking water
Pharmaceuticals	**Personal care products**	**Pharmaceuticals**	**Pharmaceuticals**
Analgesics:	Antioxidant:	Analgesics:	Analgesics:
acetaminophen	butylated hydroxy-	diclofenac	diclofenac
diclofenac	anisole (BHA)	ketoprofen	Antiepileptic drugs:
ibuprofen	Fragrances:	naproxen	carbamazepine
ketoprofen	galaxolide (HHCB)	Antiepileptic drugs:	primidone
naproxen	tonalide (AHTN)	carbamazepine	Contrast media:
Antibiotics:	Surfactants:	primidone	diatrizoate
chlorotetracycline	4-nonyl-phenol	Contrast media:	Lipid regulators:
ciprofloxacin	**Veterinary**	diatrizoate	bezafibrate
clarithromycin	**pharmaceuticals**	iopromide	clofibric acid
erythromycin-H_2O	enrofloxacin	Lipid regulators:	**Personal care products**
lincomycin	sulfadimethoxine	bezafibrate	Antioxidant:
roxithromycin	tilmicosin	clofibric acid	butylated hydroxy-
sulfadiazine	tylosin	Muscle relaxant:	anisole (BHA)
sulfadimethoxine	**Steroids**	diazepam	Fragrances:
sulfamethazine	Androgen:	**Personal care products**	galaxolide (HHCB)
sulfamethoxazole	testosterone	Fragrances:	tonalide (AHTN)
trimethoprim	Estrogens	galaxolide (HHCB)	Surfactants:
Antiepileptic drugs:	17α-estradiol	tonalide (AHTN)	4-nonyl-phenol
carbamazepine	17α-ethinyl-estradiol	Surfactants:	**Steroids**
primidone	(EE2)	4-nonyl-phenol	Estrogens
Antihypertensive:	17β-estradiol (E2)	**Veterinary**	17α-estradiol
diltiazem	estrone (E1)	**pharmaceuticals**	17α-ethinyl-estradiol
Antimicrobials:	**Other**	enrofloxacin	(EE2)
triclocarban	Caffeine and	sulfadimethoxine	17β-estradiol (E2)
triclosan	metabolite:	tilmicosin	estrone (E1)
β-blockers:	caffeine	tylosin	**Other**
atenolol	1,7-dimethyl-	**Steroids**	Flame retardants:
metoprolol	xanthine	Estrogens	tris-(2-chloro-ethyl)-
propranolol	Flame Retardants:	17α-estradiol	phosphate (TCEP)
Contrast media:	tris-(2-chloro-ethyl)-	17β-estradiol (E2)	tris-(2-chloro-
diatrizoate	phosphate (TCEP)	estrone (E1)	isopropyl)-
iopromide	tris-(2-chloro-	**Other**	phosphate (TCPP)
Lipid regulators:	isopropyl)-	Flame retardants:	Plasticizers:
bezafibrate	phosphate (TCPP)	tris-(2-chloro-ethyl)-	bisphenol A (BPA)
clofibric acid	Insecticide:	phosphate (TCEP)	di-n-butyl phthalate
gemfibrozil	n,n-Diethyl-meta-	tris-(2-chloro-	(DBP)
Muscle relaxant:	toluamide (DEET)	isopropyl)-	Stimulant:
diazepam	Nicotine metabolite:	phosphate (TCPP)	caffeine
	cotinine	Plasticizers:	
	Plasticizers:	bisphenol A (BPA)	
	bisphenol A (BPA)	Stimulant:	
	di-n-butyl phthalate	caffeine	
	(DBP)		

because it is difficult to explain the health risks of these chemicals to the public. The challenge is compounded by the ubiquitous nature of these chemicals, the inability to effectively control their introduction into wastewater treatment systems, the ability to detect some of them in water at extremely low concentrations even after conventional treatment, advanced treatment, or SAT, and a lack of information about their health effects. The detection of chemicals does not imply toxicity, but without a well-conducted toxicological assessment, these questions go unanswered—allowing for great speculation and possibly unfounded health concerns.

Some research has been conducted on human health risks related to microconstituents. One study found that environmental residues containing 17-alpha-ethinylestradiol, a key ingredient in birth control pills, was a negligible risk to humans (Christensen, 1998). Human health risk assessments for 26 active pharmaceuticals and their metabolites indicated that no appreciable risk existed when trace concentrations of these chemicals were present in surface and drinking water (Schwab et al., 2005). This information has been used to estimate both "no effect" concentrations and the number of liters of water an individual would have to consume to see an effect (Table 3.6).

More data on acceptable daily intake or safe concentrations of other EDCs and personal care products will be available in a report to be published by the American Water Works Association Research Foundation (AWWARF 308/04-003) later in 2008. This study also compares the relative estrogenicity of wastewater and a variety of food products (Table 3.7 and Figure 3.1). Some wastewaters have

TABLE 3.6 Pharmaceutical risk perspective (Snyder et al., 2005, developed from research funded by the AWWA Research Foundation. Printed with permission).

Compound	Maximum concentration in surface water (ng/L)	Predicted no effect concentration (ng/L)	Water (L/d)
Acetaminophen	10,000	5,000,000	2380
Codeine	1000	290,000	140
Gemfibrozil	790	800,000	4873
Ibuprofen	1000	16,000,000	7700
Oxytetracycline	340	4,400,000	6176
Sulfamethoxazole	1900	19,000,000	4789

TABLE 3.7 Soy sauce versus wastewater estrogenicity (EEq) (Snyder., 2007a, developed from research funded by the AWWA Research Foundation. Printed with permission).

Soy sauce		Raw wastewater		Wastewater effluent	
Kikkoman	147	WWTP-1	70	WWTP-1	4.6
Tabasco	257	WWTP-2	41	WWTP-2	0.05
Kimlan	70	WWTP-3	53	WWTP-3	0.61
La Choy	14				

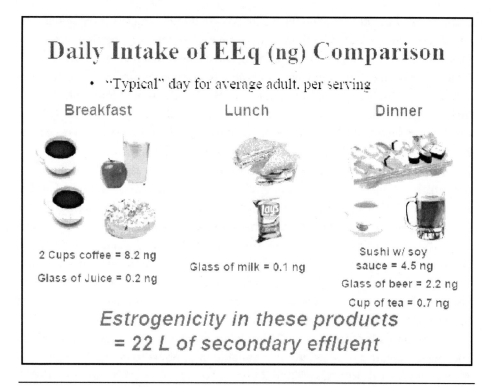

FIGURE 3.1   Daily intake of EEq via diet (Snyder, 2007a, developed from research funded by the AWWA Research Foundation. Printed with permission.).

low 17-beta-estradiol equivalents (EEq), and many dietary items had EEq that were orders of magnitude higher than those in wastewater (Snyder, 2007a).

2.4   Health-Effects Studies

Over the past 30 years, a number of studies have specifically evaluated the public heath implications of direct and indirect potable reuse. Most sought to analyze and compare the toxicological properties of reclaimed water to those of drinking water; some included epidemiological components. A summary of seven projects and related studies is presented in Table 3.8; it includes one direct potable reuse project. While these studies all had strengths and limitations, and do not necessarily apply to indirect potable reuse projects in general (Crook et al., 1999), they provided useful information for project sponsors. The study results for the Montebello Forebay Groundwater Recharge project, for example, were reviewed by an expert panel (State of California, 1987) and used to increase the amount of reclaimed water authorized for recharging groundwater. The National Research Council has recommended a number of fish-testing approaches to continuously test the toxicity of reclaimed water in indirect potable reuse projects (NRC, 1998).

TABLE 3.8   Summary of health effects studies (NRC, 1998, with updates).*

Project	Types of water studied	Health-effect data
Montebello Forebay Groundwater Recharge Study, Los Angeles County, CA (Nellor, et al., 1984; Sloss et al., 1996)	Disinfected and filtered secondary effluent, stormwater, and imported river water for replenishment; also recovered groundwater	Toxicology testing: Ames *Salmonella* test and mammalian cell transformation assay. 10,000 to 20,000× organic concentrates used in Ames test, mammalian cell assays, and subsequent chemical identification. Samples were collected from the late 1970s to the early 1980s. The level of mutagenic activity (in decreasing order) was storm runoff > dry weather runoff > reclaimed water > groundwater > imported water. No relation was observed between percent reclaimed water in wells and observed mutagenicity of residues isolated from wells. Follow-up toxicity testing of reclaimed water residues in the mid-1990s showed no Ames test response, while preserved residues from the earlier testing still showed a response, indicating that the character of the reclaimed water has changed over time, perhaps as a result of increased source-control activities.
		Epidemiology: The geographical comparison studies looked at health outcomes for 900,000 people received some reclaimed water in their household water supplies in comparison to 700,000 people in a control population. Mortality, morbidity, and cancer incidence were evaluated for 1962–1980 and 1987–1991, adverse birth outcomes for 1982–1993. The results from these studies have found that after almost 30 years of groundwater recharge, there is no association between reclaimed water and higher rates of cancer, mortality, infectious disease, or adverse birth outcomes. A household survey (women) found no elevated levels of specific illness or other differences in measures of general health.
Denver Potable Water Reuse Demonstration Project (Lauer and Rogers., 1996)	Advanced wastewater treatment (AWT) effluent (with ultrafiltration or reverse osmosis) and finished drinking water (current supply)	Toxicology testing: 150 to 500× organic residue concentrates used in 2-year *in vivo* chronic/carcinogenicity study in rats and mice and reproductive/teratology study in rats. No treatment-related effects observed.
Tampa Water Resource Recovery Project (CH2M Hill, 1993, Pereira et al., undated)	AWT effluent (using GAC and ozone disinfection) and Hillsborough River water using ozone disinfection (current drinking water supply)	Toxicology testing: Up to 1000× organic concentrates used in Ames *Salmonella*, micronucleus, and sister chromatid exchange tests in three dose levels up to 1000× concentrates. No mutagenic activity was observed in any of the samples. *In vivo* testing included mouse skin initiation, strain A mouse lung adenoma, 90-day subchronic assay on mice and rats, and reproductive study on mice. All tests were negative, except for some fetal toxicity exhibited in rats, but not mice, for the AWT sample.

TABLE 3.8   Summary of health effects studies (NRC, 1998, with updates) (*continued*).

Project	Types of water studied	Health-effect data
Total Resource Recovery Project, City of San Diego (Western Consortium for Public Health, 1996)	AWT effluent (reverse osmosis and GAC) and Miramar raw reservoir water (current drinking water supply)	Toxicology testing: 150–600× organic concentrates used in Ames *Salmonella* test, micronucleus, 6-thoguanine resistance, and mammalian cell transformation. The Ames test showed some mutagenic activity, but reclaimed water was less active than drinking water. The micronucleus test showed positive results only at the high (600×) doses for both treatments. *In vivo* fish biomonitoring (28-day bioaccumulation and swimming tests) showed no positive results.

Epidemiology: baseline reproductive health and vital statistics.

Neural tube defects study: No estimated health risk from chemicals identified based on use of reference doses and cancer potencies. |
| Potomac Estuary Experimental Wastewater Treatment Plant (James M. Montgomery, Inc., 1983) | 1:1 blend of estuary water and nitrified secondary effluent, AWT effluent (filtration and GAC), and finished drinking waters from three water treatment plants (current supplies) | Toxicology testing: 150× organic concentrates used in Ames *Salmonella* and mammalian cell transformation tests. Results showed low levels of mutagenic activity in the Ames test, with AWT water exhibiting less activity than finished drinking water. The cell-transformation test showed a small number of positive samples with no difference between AWT water and finished drinking water. |
| Windhoek, South Africa—direct reuse (Isaacson and Sayed, 1988) | AWT effluent (sand filtration, GAC) | Toxicological testing: Ames test, urease enzyme activity, and bacterial growth inhibition. *In vivo* tests include water flea lethality and fish biomonitoring (guppy breathing rhythm).

Epidemiology: Study (1976–1983) of cases of diarrheal diseases, jaundice, and deaths. No relationship to drinking water source were found. Because of Windhoek's unique environment and demographics, these results cannot be extrapolated to other populations in industrial countries. |
| Singapore Water Reclamation Study (Khan and Roser, 2007) | AWT effluent (microfiltration, reverse osmosis, UV irradiation) and untreated reservoir water | Toxicology testing: Japanese medaka fish (*Oryzias latipes*) testing over a 12-month period with two generations of fish showed no evidence of carcinogenic or estrogenic effects in AWT effluent; however, the study was repeated owing to design deficiencies. The repeated fish study was completed in 2003 and confirmed the findings of no estrogenic or carcinogenic effects.

Groups of mouse strain (B6C3F1) fed 150× and 500× concentrates of AWT effluent and untreated reservoir water over 2 years. The results presented to an expert panel indicated that exposure to concentrated AWT effluent did not cause any tissue abnormalities or health effects. |

*AWT = advanced waste treatment; GAC = granular activated carbon.

2.5   Methods To Assess Health Effects and Relative Risk

Indirect potable reuse sponsors may wish to undertake studies to assess the potential health effects or relative risks of proposed or existing projects. The following methods can be used to determine possible public health effects, establish regulatory requirements, develop appropriate treatment approaches, and forge public support for water reclamation projects. Each has various strengths and weaknesses in terms of practical application for indirect potable reuse projects (NRC, 1998).

2.5.1   Epidemiologic Investigations

Epidemiologic studies may provide useful information about whether reclaimed water is associated with any adverse effects on human health. They evaluate the relationship between an environmental pollutant and human health using data to characterize exposures to the pollutant (e.g., environmental concentrations, the probability and characteristics of human exposure, and the distributions of internal doses) and differences in the health of exposed people. There are two types of environmental epidemiology studies—descriptive and analytic—and they are not mutually exclusive. Typically, descriptive studies are most useful for generating hypotheses, and analytic studies are most useful for testing hypotheses, although each can be used for both purposes. Descriptive studies can be less definitive, but are typically less expensive and time consuming. Analytic studies are based on more detailed data on individuals that can be used to control for confounding factors; they are typically more costly and labor-intensive. Information from medical records, clinical or laboratory investigations, questionnaire results, or direct measures or estimates of exposures may allow analytic studies to explore hypotheses about suspected causes of disease or identify and measure risk factors that increase the likelihood of a given disease.

Epidemiologic studies of chemical constituents focus on their possible adverse effects on human health resulting from synergistic interaction in the human system during years of low-level exposure. Studying the health of residents in an area receiving reclaimed water allows observation of the effects of past water quality in the absence of complete information. Observing human health via epidemiologic studies is one way—albeit with limitations—to estimate whether an effect on human health has occurred because of long-term ingestion of reclaimed water.

Epidemiologic studies conducted specifically for indirect potable reuse fall into two categories: those involving reclaimed water that has been intentionally reused and those involving wastewater that has been reused unintentionally (Table 3.8). Although there are few studies of health effects related to planned indirect potable reuse, numerous studies have been conducted in the last 15 years that investigated areas receiving chlorinated surface water. These areas typically were communities that receive surface water downstream of industrial and municipal wastewater outfalls.

2.5.1.1   Ecologic Studies

Ecologic studies are often used in epidemiology as an initial look at the health effects of an environmental exposure. These studies are less expensive and time consuming than others, mainly because they rely on available morbidity, mortality, and census data. They allow more people to be included in an evaluation, while individual level studies would be limited to a much smaller sample size.

Ecologic studies rely on exposure and outcome data for groups rather than individuals. The diseased persons in a given ecologic study may not be the most exposed individuals, but this cannot be determined. Nor is information on important risk factors (e.g., smoking, alcohol consumption, and occupational/environmental exposure that might affect disease incidence) typically available or controllable in the analysis. Such factors are assumed to be equal, on average, for the exposed and control groups. If this is not true, the results may be attributable to these differences or other uncontrolled factors.

Population migration in and out of the study areas can be problematic in a chronic disease study in which years or decades may lapse between exposure and visible symptoms. Persons who move out will no longer be counted in the exposed group in the analysis, which will not bias estimates of exposure effects (assuming out-migration is independent of disease status) but does reduce statistical power by reducing the sample size of exposed persons (Hatch et al., 1990). Including people who recently moved to an area, however, will result in exposure misclassification and weaken estimates of the effect (Polissar, 1980). Another factor that affects exposure estimates is the prevalent use of bottled water.

It is important that the timing of exposure relative to the occurrence of disease be considered when interpreting results. The induction (latent) period may be long for many cancers but presumably short for infectious diseases. When studying cancer rates, several assumptions should be made about the relationship between the exposure and disease:

- One or more chemical agents in reclaimed water are carcinogenic and do not occur in other water sources;
- The chemical(s) in reclaimed water initiate or promote the carcinogenic process; and
- No matter when the "critical" exposure occurs, the resulting cancers occur during the study.

Although there is general agreement that the minimum latency period for many cancers is 15 or more years, assumptions about the latent period of cancer are not easy to prove or disprove because the role of chemical exposures and their timing in the induction of cancer is not well understood (Moolgavkar, 1994). One way to account for timing issues is to study birth outcomes, where the exposure period is shorter; however, these studies have limitations because of other factors (e.g., prenatal care) that can significantly influence effects (Sloss et al., 1999).

So, ecologic study results should be interpreted cautiously. Some of these limitations have been offset by improved environmental-exposure databases and the increased availability of sophisticated tools (e.g., geographic information systems). Ecologic studies become more valuable as the methods for estimating exposure improve, particularly if one can develop exposure gradients or can control or understand relevant confounding variables. Ecologic studies also can use some of the tools typically applied to analytical epidemiology (e.g., surveys) to gather information about confounding factors related to exposure.

2.5.1.2 Surveillance Studies

Some experts may not consider surveillance studies to be traditional epidemiologic studies, but they share some of the characteristics of descriptive epidemiology. These studies look at health outcomes or indicators for population groups that are assumed to be exposed or not exposed. No numerical or narrative exposure data is developed. In this type of study, for example, one would compare a community that uses recycled water with one that does not to see if there are differences in various health outcomes or indicators. Health outcomes could be obtained from typical disease databases, but health indicators (e.g., school absenteeism, sales of anti-diarrheal medications, and nurse hotline calls) would have to be obtained directly from the communities via surveys or other data-gathering methods.

Surveillance studies have the same confounding-factor and exposure-definition limitations associated with ecologic studies. In addition, the reliance on a dichotomous exposure (e.g., yes or no exposure determinations) probably may make it more difficult to detect an effect. Large samples would be needed for a study to have ample statistical power. It also may be difficult to identify large-enough communities with similar demographic and micro-environmental characteristics to allow for meaningful comparisons. Finally, collecting information on health indicators is likely to be difficult, expensive, and possibly of limited utility (i.e., the information may be inaccurate or unrelated to the hypothesis being tested). For example, sales of anti-diarrheal medication may result from food poisoning or stomach flu, not an environmental exposure. These studies' true value in understanding causal associations is unknown.

2.5.1.3 Case-Control Studies

Case–control studies compare the exposures of individuals with a specific adverse effect or disease (cases) with the exposures of people without it (controls). Both groups typically come from the same population. Researchers study the relationship between exposure and disease by comparing the two groups' exposure frequencies. These studies typically depend on the collection of retrospective data and may suffer from recall bias (i.e., the tendency of people with a disease to remember putative causes more readily than those without it). However, it is often possible in a case–control study to collect histories of exposure to many factors and control for confounding ones more efficiently than in a large cohort study.

2.5.1.4   Cohort Studies

Cohort studies identify a group of persons (called a *cohort*) and evaluate associations between the cohort's exposure(s) to something and one or more health outcomes over time. These studies are either retrospective or prospective; each type has advantages and disadvantages. A retrospective (historical) cohort study relates a complete set of observed outcomes in a defined population to earlier exposures; data on both exposure and outcome must be available when the study is undertaken.

In prospective cohort studies, researchers directly measure current exposure and follow individuals. These studies may have more-accurate measurements, but suffer from loss of subjects to follow-up or bias when ascertaining endpoints. Also, researchers may have to wait for many years for sufficient outcomes to occur, or even for the follow-up time to exceed the latent period between exposure and effect.

Cohort studies can use questionnaires or laboratory tests to measure both exposure and outcome. Unlike case-control studies, multiple outcomes can be evaluated simultaneously in relation to the exposure data. However, the power to test associations will depend on the frequencies of the different outcomes considered, which in turn depend on the number of persons followed.

One type of cohort study is designed to correlate time trends in outcome measures and environmental exposures. Such studies can be divided into three broad classes:

- Those in which the outcome is estimated or measured relatively few times (e.g., annually);
- Those in which outcome variables are linked to episodic variations or short exposures; and
- Those in which long-term trends in measures or estimates of health outcomes are linked with variations in monitored or estimated exposures (both exposure and outcome measures are collected for months or years).

Short-term fluctuations in those outcomes are correlated with short-term variations in environmental exposures. In most of these types of studies, the multi-factorial nature of the outcome means that the explanatory power of each environmental variable is typically small, so relatively large samples and careful modeling are needed to avoid potential confounding.

2.5.2   Chemical Risk Assessment

The goal of risk assessment is to estimate the severity of harm to human health or the environment occurring from exposure to a risk agent (Cohrssen and Covello, 1989). Via risk assessments, researchers seek to understand the fundamental processes that underlie human health problems caused by pollutants in the environment. Risk assessments address questions of exposure to chemical or biological agents in the environment and any associated adverse outcomes (e.g., cancer, birth defects, developmental disorders, and other serious health problems). In ecological risk assessments, scientists address concerns about the effects of pollution on organisms and ecosystems.

There are four steps in risk assessment: hazard identification, dose–response assessment, exposure assessment, and risk characterization. In each step, scientists address key questions with the goal of completely understanding a hazard's seriousness and scope. The U.S. Environmental Protection Agency has developed a series of risk assessment guidelines for various endpoints (e.g., carcinogenicity, neurotoxicity, reproductive toxicity, and mutagenicity), as well as guidelines on dealing with exposure assessment and chemical mixtures. Many states also have developed risk assessment approaches for setting standards and regulating contaminants.

In the hazard-identification step, researchers evaluate the types of health effects that the contaminant of concern can cause. Such effects may range from subtle, reversible physiological changes to cancer. Most are chronic and occur only after long exposure to the contaminant. This step establishes the contaminant's health effect of concern and includes discussions on the data types, quality, and uncertainties in the evaluation.

In the dose–response assessment step, researchers quantify the magnitude of the health effect with respect to exposure. For example, a *cancer-slope factor* allows researchers to estimate the probability of cancer occurring based on contaminant exposure. The *reference dose* is a "bright line" exposure level below which no adverse health effect is believed to occur. Used for noncarcinogens, this dose cannot be used to estimate probabilities of risk. Dose–response values include a number of conservative health assumptions, so they may yield an upper bound to risk estimates when the true risk is probably lower or zero.

In the exposure assessment step, researchers provide a site-specific description of the plausible amounts of a contaminant that people can receive. This assessment takes into consideration the locations and amounts of the contaminant; its movement and attenuation through the environment; the nature, routes, and frequency of possible human contact with the contaminant; and the numbers and types of people exposed. This step should include a quantitative description of the range of exposure to individuals and populations and the relevant data uncertainties and unknowns.

In the risk characterization step, researchers combine the toxicological and exposure analyses to describe the nature, magnitude, and significance of any health risks. The quantitative expression of risk can include both the average risk and the range of risks, based on the range of exposures anticipated. If the exposure assessment indicates only short-term, low exposures to contamination, health effects from contaminants requiring chronic, long-term exposures are likely to be minimal or nonexistent. Emphasis may be placed on contaminants that can cause acute health effects from short-term exposures.

There are many uncertainties in risk assessments because of the limitations in available data and the complex interactions between the sources and environmental concentrations of contaminants, the dose received at the site in a person where the effect is induced, and variability in people's responses. These uncertainties result in the use of default assumptions, simplified approaches, and uncertainty factors. Though designed to protect human health, assumptions and uncertainty factors may overestimate the risk and result in overly stringent standards and unnecessarily burdensome costs. Conversely, oversimplifying the risk assessment may underestimate risks, particularly for children or the elderly.

To overcome these challenges, U.S. EPA and others are working to develop and improve the models, assumptions, and extrapolations used in risk assessments. Recent work in this area includes an evaluation of the benchmark-dose approach for non-cancer risk assessments and the additivity-hazard-index approach for assessing risks posed by mixtures of chemicals.

For indirect potable reuse projects, quantitative risk assessments can provide useful information for decision makers. Consider drinking water MCLs, for example, which are designed to provide safe drinking water to consumers. They consist of two parts: goals and numeric standards. The MCL goals, which are unenforceable, are based on risk assessment at levels considered to be risk-free. The goal for carcinogens is zero; the goal for noncarcinogens is a zero-risk level based on the reference dose. Enforceable numeric standards, however, must consider several feasibility issues, including availability of analytical methods and treatment techniques, so zero risk cannot be obtained. The resulting numeric MCL is ultimately set at a level considered to be acceptably safe.

It is also possible to conduct a quantitative relative risk assessment for indirect potable reuse projects. This type of evaluation does not assess the absolute risk from ingesting water at the tap, but relative risk based on water quality comparisons. This approach eliminates much of the uncertainty associated with the exposure-assessment elements of standard quantitative risk assessments. Estimating situational exposure can create a high degree of uncertainty because there is no direct method of determining the uptake of chemicals in the study population's drinking water supply that originated in reclaimed water. The approach also limits the effects of many confounding factors (e.g., bottled water use, smoking, and diet) that affect exposure assessment. A quantitative relative risk assessment was conducted for the Orange County Water District's Groundwater Replenishment System (EOA, 2000). The results showed that the reclaimed water produced by the project's advanced treatment facility would be safe for consumers and actually improve groundwater quality.

2.5.3 *Microbial Risk Assessment*

Microbial risk assessment is a process that evaluates the likelihood of adverse human health effects occurring after exposure to pathogens or a medium in which pathogens are present. To date, most microbial risk assessments have used a framework originally developed for chemicals. Except for Florida, no reclaimed water standards or guidelines in the United States are based on risk assessments using microorganism infectivity models. Florida has risk assessment-based guidelines for *Giardia* and *Cryptosporidium*. These guidelines are included in a Florida Department of Environmental Protection form that is incorporated by reference into Florida's reuse rules at Chapter 62-610, F.A.C. (FDEP, 2006).

Quantitative methods for characterizing human health risks associated with exposure to pathogens began to appear in the literature in the 1970s (Fuhs, 1975; Cooper et al., 1986; Haas, 1983; Olivieri et al., 1986). Since then, microbial risk assessments have been conducted based on the assumptions that risk is manifest at an individual level and that the number of individuals susceptible to infection does not vary over time (i.e., is static) (Eisenberg et al., 2002). Examples of these types of assessments for waterborne pathogens are available in the scientific literature (Regli et al., 1991; Rose et al., 1991; Gerba et al., 1996; Mena et al., 2003). Some microbial risk assessments also focused on reclaimed water (Asano and Sakaji, 1990; Rose and Carnahan, 1992; Rose and Gerba, 1991; Tanaka et al., 1998).

Risk analyses require researchers to make several assumptions [e.g., the minimum infective dose of selected pathogens; concentration of pathogens in reclaimed water; quantity of reclaimed water (or pathogens) ingested, inhaled, or otherwise contacted by humans; and the probability of infection based on infectivity models]. Microbial risk assessment models vary primarily in how they address the unique properties of an infectious-disease-transmission system (Soller, 2006). The fundamental difference between these models is that static ones do not account for the properties that are unique to a dynamic infectious disease process.

In static models, risk focuses on the probability of an individual becoming infected or diseased as a result of one exposure event. Secondary transmission and immunity are assumed to be negligible or to cancel each other out. Static models use different epidemiologic states: a susceptible state, an infected state, and/or a diseased state. The probability that a susceptible individual will become infected when exposed to a pathogen from an environmental source depends on the number of organisms that enter the host, the host's ability to inactivate these organisms, the number of organisms that can withstand the host's local immune defenses, adhere to mucosal surfaces, and multiply to infect the host, and variation in pathogen virulence and host susceptibility (Eisenberg et al., 1996; Eisenberg et al., 2002). To estimate the number of infected/diseased individuals, researchers often multiply the probability of infection by a morbidity factor.

In a dynamic risk assessment model, the population is divided into groups of epidemiological states. Individuals move from state to state, with only a portion of the population in a susceptible state at any time, and only susceptible individuals can become infected or diseased via exposure to microorganisms. The probability that they become susceptible depends on the infective dose and infectivity of the pathogen to which they are exposed, as well as the number of infected/diseased individuals with whom they come in contact.

Microbial risk assessments can provide valuable health risk information but can have limitations for indirect potable reuse projects. Such limitations include a general lack of pathogen-specific data or, if data are available, what to do with results below detection levels—as would be expected for high-quality reclaimed water used in indirect potable reuse projects.

3.0   WATER QUALITY ISSUES

3.1   Water Quality Standards

The goal of the Clean Water Act (CWA) is to restore and maintain the chemical, physical, and biological integrity of the nation's waters. Its interim goal is to provide, wherever attainable, for recreation and the protection and propagation of fish, shellfish, and wildlife. States translated these goals into measurable water quality standards (WQS) for surface waters and groundwaters, subject to U.S. EPA approval. Each water quality standard consists of a designated use (e.g., drinking, swimming, or protecting aquatic life) and the numeric or narrative criteria that are proxies for these uses (e.g., bacterial or dissolved oxygen concentration criteria). It also includes general provisions that address implementation issues (e.g., low flows, variances from standards, and mixing zones).

States assign WQS to specific waterbodies. Some apply them statewide, with little attention to site-specific conditions (e.g., whether the use really exists or is attainable for a given waterbody). If state-conducted analyses show that the WQS assigned to a waterbody are inappropriate or unattainable, regulations allow for site-specific modifications, which must be approved by U.S. EPA.

Some states make distinctions on where drinking water uses and, hence, the applicable water quality criteria apply. In other words, some states establish a designated drinking water use for all surface waters, while others limit the drinking water use to the points where water is treated. States also typically incorporate drinking water MCLs into their WQS to protect drinking water sources (e.g., groundwater).

State criteria for toxic pollutants typically are based on U.S. EPA's recommended criteria for protecting aquatic life and human health via priority pollutants—a list of chemicals considered to be most important to control under the CWA (based on the outcome of a lawsuit with the National Resources Defense Council) (U.S. EPA, 2006b). The human health criteria are established either for consumption of water and organisms, or consumption of organisms only. Sometimes human health criteria are more stringent than drinking water MCLs because the *de minimis* risk level for human health criteria is established at 10^{-6} (versus 10^{-4} for some MCLs) and cannot take into consideration cost or technological feasibility, as is allowed under the Safe Drinking Water Act. Examples of chemicals with more stringent criteria are presented in Table 3.9.

As part of its WQS, each state is must adopt an antidegradation policy consistent with U.S. EPA's antidegradation regulations and identify the methods it will use to implement the policy. These policies address how to maintain high-quality surface waters and groundwaters, and in some cases, conditions that would allow for lowering water quality. These conditions often are tied to ensuring that designated uses are protected, are in the state's public interest, and/or apply the best practical treatment and control.

Until recently, the application of antidegradation policies has been a "sleeping giant"—rarely used or perceived as an obstacle to reuse projects. Now, however, antidegradation has become more prominent. Some states will forbid a water reclamation project to increase any water quality constituents above natural concentrations even if the waterbody will still meet state water quality and health standards. This interpretation has extended to disallowing a project or requiring more treatment if it results in chemical levels that could pose a *de minimis*

TABLE 3.9 Comparison of selected CWA human-health criteria to drinking water MCLs (U.S. EPA, 2006b, 2006c).

Compound	Clean Water Act Human-health criteria (water and organisms)	U.S. EPA MCLs (μg/L)
Total THMs	—	80
Bromoform	4.3	—
Chlorodibromomethane	0.40	—
Chlorform	Reserved	—
Dichlorobromomethane	0.55	—
Benzene	2.2	5
Bis(2-ethylhexyl)phthalate	1.2	6
Carbon tetrachloride	0.23	5
Chlordane	0.00080	2
1,2-Dichloroethane	0.38	5
Tetrachloroethylene	0.69	5
Trichloroethylene	2.5	5
Toxaphene	0.00028	3

health risk. The key pollutants that have been targeted via antidegradation policies are salts, nitrogen compounds, and, in a few cases, microconstituents.

Antidegradation is an important consideration for indirect potable reuse projects. An example is Gwinnett County's proposed discharge of reclaimed water into Lake Lanier near Atlanta, Georgia (City of San Diego, 2006). The proposed treatment train consisted of secondary treatment for nutrient removal, membrane filtration, multimedia and activated carbon filters, and ozone disinfection. A lawsuit was filed to prevent the discharge. It was not based on indirect potable reuse concerns but rather on concerns that the reclaimed water discharged into Lake Lanier would contain nutrients that might encourage the growth of algae and other aquatic plants. The Georgia Supreme Court ruled that Gwinnett County's discharge permit would not protect Lake Lanier's water quality, stating that "the clear and unambiguous language of Georgia's antidegradation rules require the permittee to use the 'highest and best (level of treatment) practicable under existing technology.'"

An example in which antidegradation policies were successfully applied to facilitate water reclamation is the 2004 update of the TDS and nitrogen management plan in Southern California's Santa Ana River Basin Water Quality Control Plan (SARWQCB, 2004). This stakeholder-led effort resulted in the adoption of alternative, less-stringent TDS and nitrate–nitrogen water quality criteria for specific groundwater management zones and surface waters. A number of agencies made appropriate designated use protection and "maximum benefit" demonstrations to justify alternative maximum-benefit water quality criteria for a number of groundwater management zones. These maximum-benefit proposals entailed agency commitments to implement specific projects and programs to address salt problems in the region's groundwater, including constructing brine lines and groundwater desalters; implementing programs to enhance the recharge of high-quality stormwater and imported water, where available; and re-injection of reclaimed water to maintain saltwater-intrusion barriers in coastal areas. To address circumstances that might impede or preclude these commitments, the revised basin plan included both "antidegradation" and "maximum benefit" objectives for the groundwater management zones. The antidegradation objectives are more stringent than the maximum-benefit objectives. As long as the agencies' commitments are met, then they have demonstrated maximum benefit, and the maximum-benefit objectives in the 2004 basin plan apply for regulatory purposes. However, if these commitments are not being met, then maximum benefit is not demonstrated and the antidegradation objectives will apply.

Another related challenge is the allocation of assimilative capacity. *Assimilative capacity* is the amount of a contaminant that can be discharged to a specific waterbody without exceeding water quality standards or criteria. For groundwater, this would be the difference between a contaminant's background concentration and its water quality criterion. When assessing permit limits for a reclamation project, a state may consider the groundwater basin's available as-similative capacity but may not be willing to authorize the use of some or all of it for that project. If assimilative capacity is not provided, it can result in stringent permit requirements that could affect the control measures needed for project approval, discouraging their implementation. It also can lead to situations in which projects are not allowed to proceed.

3.2 Discharge Permits

As authorized by the CWA, the National Pollutant Discharge Elimination System (NPDES) permit program controls water pollution by regulating point sources that discharge pollutants into U.S. waters. Industrial, municipal, and other facilities must obtain NPDES permits if their discharges go directly into "waters of the U.S." (e.g., surface waters and natural wetlands), which are so designated by the U.S. Army Corps of Engineers. In most cases, the NPDES permit program is administered by an authorized state, but sometimes the program is administered by the regional U.S. EPA office. Discharges to "waters of the state or land" are regulated via state-issued permits. Indirect potable reuse projects may be regulated under either NPDES or state permits, depending on how the water is introduced to a drinking water supply, if the receiving water meets the definition of a U.S. water, and state policy and practice.

National Pollutant Discharge Elimination System permits contain technology-based limits pursuant to the CWA. For municipal wastewater dischargers, the limits are based on secondary treatment [e.g., 5-day BOD (BOD_5) and total suspended solids (TSS) limits and specified removals]. For industrial dischargers, whether direct (discharging to U.S. waters) or indirect (discharging to municipal wastewater collection systems), the limits are based on the U.S. EPA's effluent guidelines for conventional and toxic pollutants. These guidelines are established based on the amount of pollutant reduction that an industrial category can attain by applying pollutant-control technologies. The standards for indirect dischargers also consider what wastewater treatment plants remove.

When technology-based limits in NPDES permits are insufficient to maintain WQS, the permits will include water-quality-based limits. Such limits are based on the state's criteria (numeric and narrative) and the reasonable potential for a discharge to cause or contribute to an exceedance of a WQS. This is determined based on effluent data and receiving water conditions. So, an indirect potable reuse project's NPDES permit could have more stringent limits than one without an NPDES permit, particularly if dilution is unavailable (e.g., the effluent is discharged to a dry creek or stream) or not allowed (some states disallow dilution depending on the receiving water or pollutant). Stringent limits could be an obstacle to implementing projects if more treatment is needed or if the limits cannot be attained.

This is also an important consideration for projects that incorporate wet-lands into indirect potable reuse projects. Wetlands can provide additional nutri-ent treatment and, depending on site conditions, recharge can occur *in situ* (West Palm Beach, Florida; Tres Rios Wetlands, Arizona). In other cases, wetlands are used as a pretreatment step before the water is released to spreading basins (Tucson, Arizona) or surface reservoirs (planned projects in Texas). Planners of indirect potable reuse projects may want to take credit for wetlands enhance-ment or use natural wetlands, which should be considered carefully in light of the NPDES permitting consequences. Because natural wetlands are U.S. waters, a project that discharges to a natural wetland must be permitted under the NPDES program, while a "treatment" wetland is considered part of the treat-ment process, not a U.S. water. Permitting is also a factor for recharge projects that use in-stream spreading (planned San Gabriel Valley Groundwater Recharge Project, California) rather than off-stream spreading areas (Montebello Forebay Recharge Project, California; Orange County Groundwater Replenishment Sys-tem, California). In-stream planned recharge would require an NPDES permit, while an off-stream spreading ground would not. One project team that elected to use in-stream spreading found that the resultant permit requirements in-volved more treatment, which may ultimately result in the project not being im-plemented (San Gabriel Valley Recharge Project, California). Permit requirements also apply if agencies use U.S. waters as conveyance channels to send reclaimed water to spreading grounds (Montebello Forebay Recharge Project, California) or surface reservoirs. While using natural conveyance channels offers cost sav-ings and the possibility of further treatment (via photolysis), the NPDES-permit and water-rights ramifications might make piping water directly to recharge or surface augmentation sites more cost-effective (see Total Maximum Daily Loads and Water Rights sections in this chapter).

Averaging periods for effluent limits are potentially more flexible for non-NPDES permits than for NPDES permits. One benefit is that non-NPDES per-mits can use annual averages (similar to those used for drinking water pro-grams) rather than the instantaneous, daily, or monthly averages used in NPDES permits.

As part of the NPDES program, municipal wastewater management agen-cies [or publicly owned treatment works (POTWs)] designed to treat more than 219 L/s (5 mgd) and smaller POTWs with significant industrial discharges (e.g., those subject to U.S. EPA's Effluent Guidelines) must establish local pretreat-ment programs. Such programs regulate indirect dischargers—industries and businesses that discharge to the POTW's wastewater collection system. The pro-grams, which have permitting, inspection and enforcement elements, must en-force all national pretreatment standards and requirements, as well as any more-stringent local requirements necessary to protect the POTW's site-specific conditions. Some states require all or most POTWs (e.g., except for those not in the NPDES program, with lower design flows, or with no significant industrial dischargers) to develop and implement pretreatment programs. Pretreatment programs are critical to the success of indirect potable reuse projects because they are the initial means of preventing undesirable chemicals or difficult-to-treat pollutants from entering wastewater collection systems.

National Pollutant Discharge Elimination System permits include monitoring and reporting programs for wastewater and receiving waters to determine if uses and criteria are being achieved; non-NPDES permits also include site- and use-specific monitoring requirements. National Pollutant Discharge Elimination System permits also require agencies to test effluent (and sometimes receiving waters) for acute and chronic toxicity. If toxicity is discovered, the discharger must conduct a toxicity-reduction evaluation to identify and reduce or eliminate it.

3.3   Total Maximum Daily Loads

If standards are not met in a U.S. waterbody because of a pollutant that cannot be controlled via technology- or water quality-based effluent limits, then the waterbody is listed as "impaired" for that pollutant [e.g., it is placed on a state's CWA Sec. 303(d) list of impaired waters], and a total maximum daily load (TMDL) must be developed to meet the applicable WQS. The *total maximum daily load* is a calculation of the maximum amount of a pollutant that a waterbody can receive from point and nonpoint sources and still meet the WQS with a margin of safety. These loads are translated into NPDES permit limits to ensure that discharges attain WQS.

Before a TMDL is developed, new or expanded discharges to impaired waters typically are prohibited if they contain a pollutant that could cause or contribute to the impairment. Most water quality-impairment listings and TMDLs are for nutrients, bacteria, and sediment, although more listings are including metals, salts, and organic chemicals. If agencies use waters on the 303(d) list as part of an indirect potable reuse project, they may have to provide more controls—particularly for nutrients or salts—on an unrelated project or may be subject to restrictions on project expansions without further load reductions. They also may be caught in the time gap between listing and completing a TMDL, which can complicate obtaining or revising a permit and obtaining reasonable effluent limits.

Sometimes utilities elect to switch to water reclamation to achieve TMDL wasteload allocations. Water reuse, which diverts discharges to surface waters, reduces pollutant loads to impaired surface waters. This situation could be an incentive for indirect potable reuse. Utilities considering indirect potable reuse that affect U.S. waters should be cognizant of the waterbody's impairment status and the development of TMDLs.

3.4   Enforcement

State and NPDES permit violations have both civil and criminal liability. Some states (e.g., New Jersey and California) have adopted laws specifying mandatory minimum penalties for NPDES violations of effluent limits and reporting requirements. Continuing violations of NPDES permits are also subject to third-party lawsuits under the Clean Water Act.

4.0   OTHER REGULATORY/STATUTORY ISSUES

4.1   Water Rights

Indirect potable reuse project planners and designers should be cognizant of local water rights laws and procedures because statutory requirements apply when determining who has the rights to reclaimed water used for indirect potable reuse, as well as to plans to remove or reduce effluent discharges to surface waters. Water rights law determines the extent to which an individual can use the water that runs across, underlies, or moves through the atmosphere above a person's property. Such laws are complex and can be a significant obstacle to indirect potable water reuse.

In the United States, two major water rights doctrines—riparian and prior-appropriation systems—are applied to surface water in streams (Templer, 1991). Riparian water rights are largely unregulated and unquantified and are tied to riparian land ownership. The prior-appropriation system, on the other hand, is administered by a state agency, and appropriative rights are quantified according to purpose, quantity, place, and (occasionally) time of use. In general, the differences between the common-law riparian and prior-appropriation systems also apply to groundwater. Three common-law variations have developed: the strict common-law rule (absolute ownership), the doctrine of reasonable use, and the doctrine of correlative rights. Some states have additional types of water rights classifications (e.g., prescriptive and Pueblo).

That said, determining who has rights to reclaimed water or water supplemented with reclaimed water is a complex legal issue that varies from state to state. So, indirect potable use project planners should be cognizant of relevant water-rights issues and aware of state requirements and procedures, which can range from water rights assessments (Washington) to petitions (California).

4.1.1 Texas

According to Texas law, treated wastewater that is not reused but rather discharged to a watercourse is *return flow*. The subsequent downstream diversion and use of return flows is *indirect reuse*. Indirect reuse uses a state watercourse for transportation, thereby saving the capital cost of pipelines and possibly treatment. Like direct reuse, indirect reuse ultimately reduces the amount of flow in the watercourse that is available to other water rights holders and the environment (Reuse Committee of the Texas Water Conservation Association, 2006). This effect, of course, is most evident downstream of the point where indirect reuse occurs. Upstream of this point, return flows provide some instream flow benefit.

While Texas agencies can engage in direct reuse without water-rights-permitting implications, the ability to engage in indirect reuse is less clear. A large number of water-rights applications for indirect-reuse authorization are pending before the Texas Commission on Environmental Quality, nearly all of which have been protested. It currently is unclear how Texas will permit indirect reuse of surface-water-derived effluent, if at all. Such permits could have enormous implications (e.g., who ultimately might obtain such rights, what their value will be, and how potential effects on other water users and the environment might be addressed).

4.1.2 California

Under California law, the holder of an appropriative right may change the point of diversion, place of use, or purpose of use, so long as other rights are not injured by the change. To change a water right's attribute in California, a change application must be filed with and approved by the State Water Resources Control Board. Reuse projects that involve changes in the discharge point can become embroiled in water rights issues.

Consider the example of a utility permitted to discharge effluent to a river that later proposed to reuse this effluent instead. The river's flows support both irrigation and municipal use, and the State Water Board determined that downstream third-party water-right holders already put the effluent to full beneficial use. The Board determined that diverting effluent upstream for reuse would reduce the water available to other downstream users and would not promote maximum beneficial use, but take water away from parties who were already beneficially reusing the effluent. So, the Board determined that any diversion of reclaimed water would injure all of the water-right holders and denied the diversion.

Most of California's groundwater rights are unregulated; only a few basins are adjudicated or subject to controls. The state does not have a comprehensive groundwater permit process to regulate groundwater withdrawal. Landowners overlying percolating groundwater may use it on an equal and correlative basis (i.e., all property owners above a common aquifer possess a shared right to reasonable use of it). When reclaimed water is used to recharge unadjudicated groundwater basins, water-rights conflicts arise that could derail projects.

4.2 Endangered and Threatened Species

Under the federal Endangered Species Act, an *endangered species* is a species of plant or animal that is in danger of becoming extinct throughout all or in a significant part of its range, and a *threatened species* is a species of plan or animal that is likely to become endangered soon. Indirect potable reuse project planners should be aware of any endangered or threatened species issues related to project sites or the diversion of reclaimed water.

For example, indirect potable reuse projects may increase or decrease instream flows. Long-term wastewater discharges may have created a habitat that supports endangered or threatened species. If instream flows might drop, utilities may be required to maintain minimum base flows to protect species that live in the streambed, thereby limiting the amount of reclaimed water that can be dedicated to a project. Alternatively, they may be required to supplement base flows with other sources of water, which can affect costs. If instream flows increase and could damage habitat, agencies may be required to mitigate such effects by controlling or equalizing releases or else creating new habitat.

5.0   STATE REGULATIONS AND GUIDELINES FOR INDIRECT POTABLE REUSE

For regulators, the challenge is developing indirect potable reuse criteria that ensure that high-quality potable water supplies are maintained regardless of their source. Currently, reclaimed water for indirect potable reuse is not regulated consistently nationwide. There are no federal regulations except those related to the use of injection wells. Five states have adopted indirect potable reuse regulations or guidelines; the rest approve indirect potable reuse projects on a case-by-case basis.

Imposing criteria allows project teams to plan reclamation projects based on defined regulatory requirements. Standards also imply a regulatory endorsement of indirect potable reuse that may bolster public confidence and encourage more use of reclaimed water to augment potable water supplies.

5.1   Regulations for Injection Wells

There are federal and state requirements for aquifer storage and recovery (ASR) projects and saltwater-barrier projects that use injection wells. The U.S. EPA Underground Injection Control (UIC) program categorizes injection wells into five classes, only one of which (Class V) applies to indirect potable reuse projects. Under existing federal regulations, Class V injection wells are "authorized by rule", which means they do not require a federal permit if they do not endanger underground sources of drinking water and comply with other UIC program requirements. However, states may require that specific treatment, well construction, and water quality monitoring standards be met before authorizing and permitting the injection of reclaimed water into aquifers that are or could be used for potable supply.

5.2 State Regulations/Guidelines

Five states have developed indirect potable reuse regulations or guidelines: California, Florida, Hawaii, Idaho, and Washington. Florida also has adopted regulations that address ASR projects.

5.2.1 California

Any intentional augmentation of drinking water sources with reclaimed water in California requires two state permits. A waste discharge (NPDES or non-NPDES) or reclamation permit would be required from a California Regional Water Quality Control Board based on recommendations provided by the California Department of Public Health (CDPH). The public (drinking) water systems using the affected source may need an amended water-supply permit from the CDPH to address changes in source(s).

The department adopted narrative regulations (CDPH, 2001) for surface-spreading projects using reclaimed water. These regulations require the reclaimed water quality to be protective of public health and specify that the agency will make recommendations on a case-by-case basis. When making such recommendations, the CDPH considers treatment provided, effluent quality and quantity, spreading area operations, soil characteristics, hydrogeology, residence time, and distance to withdrawal. Although the regulations are silent on direct-injection projects, in practice the same principles are used when making recommendations for approving and permitting these projects. Although no surface-augmentation projects have yet been permitted in California, it is believed that the approach proposed for the San Diego Water Repurification Project, which has been endorsed by CDPH and a panel of experts, would serve as the model for such projects. As conceived, the project would introduce reclaimed water that had undergone advanced waste treatment (ultrafiltration, reverse osmosis, and advanced oxidation with UV and hydrogen peroxide) to a water-supply reservoir, where it would mix with the indigenous water and be withdrawn for further treatment in accordance with the California Surface Water Treatment Rule before being distributed as potable water.

Over time, the CDPH developed a series of draft groundwater-recharge regulations for surface spreading and direct injection that are used as guidelines in approving and permitting projects. The current version of the proposed criteria (CDPH, 2007) relies on a combination of controls and barriers to maintain a microbiologically and chemically safe groundwater-recharge operation. These controls and barriers that have been demonstrated in existing projects (e.g., the Montebello Forebay Groundwater Recharge Project, Water Factory 21, West Basin Sea Water Barrier Conservation Project, and Chino Basin Groundwater Recharge Project). Total organic carbon is the primary means of controlling microconstituents in the reclaimed water.

The draft regulations' key elements include

- General provisions (e.g., implementing industrial source-control programs).
- Minimum treatment and disinfection requirements for reclaimed water based on a disinfected, filtered effluent:
  - Turbidity cannot exceed a daily average of 2 nephelometric turbidity units (NTU), 5 NTU for more than 5% of a 24-hour period, or 10 NTU at any time;
  - Total coliforms cannot exceed a 7-day median of 2.2 most probable number (MPN)/100 mL, 23 MPN/100 mL in one sample in any 30-day period, or more than 240 MPN/100 mL in any sample;
  - Concentration of chlorine multiplied by chlorine contact time (CT) of 450 mg-min/L.
- A minimum retention time for recharged water (6 months for surface-spreading projects and 12 months for injection projects) before being withdrawn as a source of drinking water.
- A minimum horizontal separation [150 m (500 ft) for surface-spreading projects and 600 m (2000 ft) for injection projects] between the recharge point and the withdrawal point.
- Nitrogen standards for reclaimed water.
- Reclaimed water compliance with primary and secondary MCLs (except color).
- Standards for diluent water (e.g., stormwater or surface water).
- Limits on reclaimed-water contributions [ending is required for most projects, with initial maximums for surface-spreading projects (20%) and injection projects (50%)].
- Operating requirements and other provisions for increasing reclaimed-water thresholds.
- TOC standards for reclaimed water (as a surrogate for unregulated organics).
- Operation optimization.
- Monitoring requirements for regulated constituents and microconstituents in reclaimed water, diluent water, and groundwater.
- Requirements for engineering and other reports.

Research is under way to evaluate surrogates and indicators other than TOC to control microconstituents. A WateReuse Foundation project, Development of Indicators and Surrogates for Chemical Contamination Removal during Wastewater Treatment (WRF-03-014; Drewes et al., 2007), is designed to evaluate and recommend surrogates and indicators to monitor wastewater-derived contaminants, validate analytical methods for them, and validate surrogates and indicators for assessing the performance of unit operations and treatment trains. In this study, *indicators* are quantifiable levels of individual compounds that can represent certain physiochemical and biological characteristics relevant to contaminant fate and transport during treatment or SAT. *Surrogates* are bulk parameters that can serve as performance measures for individual unit processes or operations. The goal is to develop a unique set of indicators and surrogates that can be applied to injection projects and surface-spreading projects for use in validating process performance, during startup and shakedown, and monitoring compliance during operations. At press time, the project report was expected to be available in 2008 (Drewes et al., 2007).

5.2.2 Florida

The Florida Department of Environmental Protection (FDEP) has developed comprehensive, detailed rules for reusing reclaimed water (FDEP, 2006). They include provisions for groundwater recharge, indirect potable reuse, salinity barrier systems, discharge to surface waters that are directly connected to potable groundwaters (canal recharge), and ASR. Florida's high-level disinfection requirements are required for public-access reuse systems (e.g., using reclaimed water to irrigate residential properties, golf courses, parks, schools, and edible crops; flushing toilets; fighting fires; and filling decorative water features), as well as for some groundwater-recharge and indirect potable reuse systems.

These disinfection standards were based on research documenting that a 5.0 mg/L TSS limit coupled with a "nondetect" fecal coliform limit produced a reclaimed water that is essentially pathogen free (Wellings, 1980). In any 30-day period, at least 75% of all fecal coliform observations must be nondetect and no sample can exceed 25 fecal coliforms per 100 mL. While the original work focused on viruses, a subsequent study documented the effectiveness of this disinfection level in removing or inactivating other pathogens (e.g., *Giardia*, helminths, and *Cryptosporidium*) (Rose and Carnahan, 1992). The total suspended solids limit has proven to be effective in controlling filtration to maximize disinfection efficiency. To ensure compliance with the single-sample maximum of 5.0 mg/L, filtration systems typically are operated to achieve average TSS concentrations of approximately 1 to 2 mg/L. At this TSS level, turbidities typically are less than 1.0 NTU.

Florida also defined "full-treatment disinfection," which is required for some groundwater-recharge and indirect potable reuse systems. Full-treatment disinfection is modeled on primary drinking water standards and uses total coliforms as the indicator organism. Daily sampling of total coliforms is required, and no more than one observation per month may be at detectable levels. For groundwater-recharge projects, a single-sample maximum of 4 total coliforms per 100 mL also applies.

Reclaimed water may be used to augment supplies in Class I surface waters (used for potable supplies) if it meets primary and secondary drinking water standards, a 10-mg/L total nitrogen limit (annual average), and a 3.0-mg/L TOC limit (monthly average; 5.0 mg/L single-sample maximum). The treatment facility must provide Class I reliability (U.S. EPA, 1974), filtration, and full-treatment disinfection. Reclaimed water outfalls cannot be within 150 m (500 ft) of a potable water intake.

Rapid-rate land-application systems may be used to recharge groundwaters. Nearly all freshwater groundwater in Florida is classified as *G-II*—groundwater containing 10 000 mg/L or less of total dissolved solids with a designated use of "potable supply." The regulations require secondary treatment and basic disinfection. A nitrate limit of 12 mg/L (as nitrogen, single-sample maximum) is imposed. For absorption fields, a TSS limit of 10 mg/L (single-sample maximum) is imposed to protect against formation plugging. For new systems, the initial average annual hydraulic loading rate typically is limited to 0.76 m/d (3 in./d), while for existing systems, the average annual hydraulic loading rate is limited to 0.23 m/d (9 in./d) and wetting and drying cycles must be used. A 150-m (500-ft) setback distance is required from land-application systems to potable water supply wells.

The regulations impose more stringent controls on rapid-rate systems with average annual hydraulic loading rates more than 0.23 m/d (9 in./d), continuously loaded systems (which do not feature wetting and drying cycles), and systems in unfavorable hydrogeologic conditions (karst areas and areas with clean sands over unconfined aquifers). Projects permitted under these provisions must provide secondary treatment, filtration, high-level disinfection, and Class I reliability. After filtration, TSS is limited to 5.0 mg/L (single-sample maximum). In addition, reclaimed water must meet primary and secondary drinking water standards and a total nitrogen limit of 10 mg/L (as nitrogen; annual average).

Florida's reuse rules provide for several cases of groundwater recharge by injection that are based on the receiving groundwater's TDS at the injection point. The most stringent case involves injection to potable groundwater containing 3 000 mg/L or less of TDS. In this case, treatment facilities must provide secondary treatment, filtration, full-treatment disinfection, and any other unit processes needed to meet reclaimed water limits. Class I reliability must be provided. Total suspended solids are limited to 5.0 mg/L (single-sample maximum) following filtration. The reclaimed water must meet primary and secondary drinking water standards and have a TOC of 3.0 mg/L (monthly average), with no sample exceeding 5.0 mg/L. In addition, the reclaimed water must have TOX of 0.2 mg/L (monthly average), with no sample exceeding 0.3 mg/L. Total nitrogen is limited to 10 mg/L (as nitrogen; annual average). The treatment facilities must provide multiple barriers to control pathogens and organic compounds. A 12-month pilot-testing program is required. Injection wells may not be within 152 m (500 ft) of potable water supply wells.

Recognizing that groundwaters with higher TDS concentrations typically will need more treatment upon withdrawal, the requirements for injection into groundwaters containing between 3 000 and 10 000 mg/L of TDS are less stringent than those for injection into groundwaters with less than 3 000 mg/L of TDS. The injected reclaimed water must meet primary drinking water standards, and full-treatment disinfection is required. The 10-mg/L total nitrogen limit is imposed, but TOC and TOX limits do not apply. The injected reclaimed water is not required to meet secondary drinking water standards or the state-established primary drinking water limit for sodium. A discharge zone is allowed for sodium and secondary drinking water standards, but all groundwater standards must be met at its edge. Pilot testing is not required. The discharge zone must not extend within 152 m (500 ft) of a potable water supply well.

Florida's rules provide for the use of reclaimed water to prevent or retard landward or upward intrusion of saltwater into potable-quality groundwaters. Both rapid-rate land-application systems and injection wells may be used, and specific criteria are established.

Florida's rules also address "canal recharge" systems. This concept has potential application in Southeast Florida, where an extensive network of canals delivers water to recharge the Biscayne Aquifer—a shallow, unconfined, sole-source aquifer that is the primary source of drinking water for Miami–Dade and Broward Counties. Secondary treatment, filtration, and high-level disinfection are required. The 10-mg/L total nitrogen limit applies. In addition, canal-recharge systems must be designed and operated so water entering the groundwater system meets groundwater standards (primary and secondary drinking water standards) at the entry point(s). As with any surface water discharge, all applicable surface water quality standards must be met, and the state's antidegradation policy applies. Further, to be considered "reuse," the applicant must affirmatively demonstrate the existence of a connection between surface and groundwater, the groundwater's need for more water, and the proposed discharge's ability to recharge the groundwater effectively.

Florida's rules also address reclaimed-water ASR. As noted in this rule, ASR is regarded as an attractive storage method for public-access reuse systems. The reclaimed-water ASR system must be designed and operated as an underground storage system; because the water is withdrawn for the public-access reuse system later, the ASR system is not considered "reuse" or a groundwater-recharge system. Detailed requirements for reclaimed-water ASR systems are based on the receiving groundwater's TDS concentration. Because these are storage systems, the requirements typically are less stringent than those for groundwater-recharge projects using injection. Sometimes extended discharge zones are allowed.

5.2.3 Hawaii

The State Department of Health (DOH) has issued Guidelines for the Treatment and Use of Reclaimed Water that contain requirements for both the purveyors and users of reclaimed water. The agency's intent is to incorporate the Guidelines into Chapter 11-62 of the Hawaii Administrative Rules. Groundwater recharge with reclaimed water is an allowable use per the state's guidelines, but DOH evaluates the feasibility of proposed projects on a case-by-case basis (U.S. EPA, 2004).

5.2.4 Idaho

The Idaho Department of Environmental Quality (IDEQ) has adopted regulations for using Class A reclaimed water to recharge groundwater via surface spreading or subsurface distribution (IDEQ, 2007b). Treatment requirements for Class A reclaimed water consist of oxidation, coagulation, clarification, filtration, and disinfection (or treatment by an equivalent process). Requirements for filtration approval, nutrient removal, turbidity, monitoring, reliability and redundancy, and distribution system also apply. Class A treatment systems must be pilot-tested or otherwise approved, and project teams must provide an engineering report.

The following requirements apply to Class A reclaimed water used for groundwater recharge:

- ≤ 5 mg/L of BOD_5 (monthly average), as determined via weekly composite sampling;
- $\leq 2.2/100$ mL (median) and $\leq 23/100$ mL (7-day median) of coliforms using a CT of 450 mg-min/L with a modal contact time of 90 minutes or an equivalent process that can achieve 5-log inactivation of virus;
- ≤ 2 NTU (daily arithmetic mean) and ≤ 5 NTU (at any time) of turbidity for systems using cloth, sand, or other granular media; ≤ 0.2 NTU (daily arithmetic mean) and ≤ 5 NTU (at any time) of turbidity for systems using membrane filtration; and
- ≤ 10 mg/L of total nitrogen at the compliance point (it may be lower depending on the results of any applicable nutrient–pathogen studies).

For any groundwater-recharge system, Class A reclaimed water also must meet groundwater quality standards per IDEQ's Ground Water Quality Rule (IDEQ, 2007a). Groundwater recharge sites must be at least 300 m (1 000 ft) from any downgradient drinking-water extraction well and must provide for at least 6 months' travel time in the aquifer before withdrawal. The minimum requirements for site location and aquifer storage time may be greater depending on any source-water assessment-zone studies for public drinking-water wells in the area. Owners or operators of groundwater recharge projects are required to control the ownership of the downgradient area to prohibit future wells from being drilled in the groundwater-recharge system's impact zone. IDEQ requires more permits for groundwater-injection wells.

5.2.5   Washington

The Washington State Departments of Health and Ecology (DOHC) have developed regulations for groundwater recharge via surface percolation and direct aquifer recharge (DOHC, 1997). Final standards have not been developed for recharge via surface percolation, but criteria for minimum treatment, pretreatment, reliability, and submittal of an engineering report are applied. The reclaimed water must undergo secondary treatment so BOD $\leq$ 5 mg/L and TSS $\leq$ 5 mg/L. It also must undergo another treatment step to reduce nitrogen before final discharge to groundwater. After disinfection, reclaimed water must contain < 2.2/100 mL of total coliforms (7-day average), and no more than 23/100 mL in any sample.

The requirements are more extensive for direct aquifer recharge projects. They are separated into standards for nonpotable and potable groundwaters. The following standards apply to projects affecting potable groundwater:

- All known, available, and reasonable methods of prevention, control, and treatment shall be applied to all wastewater before direct recharge;
- Reclaimed water must be an oxidized, coagulated, filtered, and disinfected wastewater that also has undergone reverse osmosis;
- Reclaimed water must meet the water quality criteria for primary contaminants (except nitrate), secondary contaminants, radionuclides, and specifically listed carcinogens; total coliforms must be < 5/100 mL in any sample; turbidity must be $\leq$ 0.1 NTU (average) and 0.5 NTU (maximum); total nitrogen must be $\leq$ 10 mg/L as nitrogen; TOC must be $\leq$ 1.0 mg/L; it also must meet any other constituent limits deemed appropriate (compliance is determined at a location immediately before injection);
- Monitoring is required for both reclaimed water and groundwater;
- Reclaimed water must be retained in the aquifer for at least 12 months before being withdrawn as a source of drinking water;
- The discharge and withdrawal points must be at least 600 m (2000 ft) apart (horizontal separation);
- The treatment processes must meet reliability requirements;
- An engineering report must be prepared; and
- A pilot study must be conducted.

5.3   What To Do in the Absence of State Regulations/Guidelines

There are proposed and existing projects in states without established indirect potable reuse regulations. The following information for two such states—Arizona and Texas—illustrates the requirements used for permitting such projects. Similar procedures could be used in other states without indirect potable reuse regulations.

5.3.1 Arizona

Arizona's water reuse regulations do not address indirect potable reuse; they specifically state that the use of reclaimed water for direct human consumption is prohibited (State of Arizona, 1991). The use of reclaimed water for groundwater recharge is regulated under statutes and administrative rules administered by the Arizona Department of Environmental Quality (ADEQ) and the Arizona Department of Water Resources (ADWR). In general, ADEQ regulates groundwater quality and ADWR manages groundwater supply. These agencies require several permits before a groundwater-recharge project is implemented.

In Arizona, a groundwater-recharge project is considered a discharge facility, which is regulated under the Aquifer Protection Permit Program administered by ADEQ. The owner or operator of a groundwater-recharge project using reclaimed water must obtain an aquifer-protection permit before any reclaimed water can be recharged (ADWR, 1995). A wastewater treatment plant that provides reclaimed water for groundwater recharge also is considered a categorical discharge facility and must obtain an aquifer-protection permit. One permit may be issued if the same applicant applies for both permits for facilities in a contiguous geographical area.

To obtain an aquifer-protection permit for a project using reclaimed water to recharge groundwater, a permit applicant must demonstrate that the project will not cause or contribute to the violation of an aquifer water quality standard and that the treatment plant providing the reclaimed water demonstrates compliance with best available demonstrated control technology (BADCT) requirements. All aquifers in Arizona currently are designated for drinking water use. Aquifer water-quality standards based on MCLs have been adopted to protect groundwater quality. Determining BADCT for an existing treatment plant involves evaluating the technical and economic feasibility of retrofitting the plant to include state-of-the-art treatment technologies. While certain treatment processes are described in the BADCT guidance document, the optimal reduction of pollutants is defined in terms of concentrations of certain indicators:

- $< 2.2/100$ mL of fecal coliforms (geometric mean);
- 1.0 NTU of turbidity;
- < 10 mg/L of nitrogen; and
- MCLs of hazardous substances or, if no MCL, an action level or concentration representing a 10^{-6} cancer risk level, whichever is lower.

In evaluating the BADCT for a new or existing treatment plant for groundwater recharge, ADEQ may take into account the recharge site's site-specific characteristics and operating processes if they reduce the concentration of pollutants before reclaimed water reaches the aquifer. So, it may be feasible to demonstrate that filtering effluent at the treatment plant is unnecessary because SAT at the recharge site will meet the 1.0 NTU turbidity limit at the applicable compliance point.

Aquifer water-quality standards are determined in the aquifer at a *point of compliance*, which is defined as a vertical plane downgradient of the facility that extends through the uppermost aquifers underlying the facility. A primary factor in determining the point of compliance for a groundwater-recharge project using reclaimed water is the reclaimed water's waste characterization. If the reclaimed water contains a hazardous substance, then the groundwater-recharge project's point of compliance must be in a pollutant management area.

A *pollutant management area* is the limit projected in the horizontal plane of the area on which pollutants are or will be placed, including "any horizontal space taken up by any liner, dike or other barrier designed to contain pollutants in the facility" (ADWR, 1995). For a groundwater-recharge project that will recharge reclaimed water containing one or more hazardous substances via surface spreading, the point of compliance will be at the boundary of the infiltration basins. If there are multiple infiltration basins, then the pollutant management area is described by an imaginary line that circumscribes all of the basins.

Any groundwater-recharge project involving the direct injection of reclaimed water into an aquifer is required to demonstrate compliance with aquifer water-quality standards at the injection point. Reclaimed water must be treated to meet drinking water standards before it could be injected directly into an aquifer.

A groundwater-recharge project that uses reclaimed water is required to obtain an underground storage facility permit from ADWR. To obtain such a permit, an applicant must demonstrate that

- The owner has the technical and financial capability to construct and operate the groundwater-recharge project;
- The maximum amount of reclaimed water that could be in storage at any one time is hydrologically feasible;
- Storing this amount of reclaimed water will not cause unreasonable harm to land or other water users in the project's impact area for the duration of the permit;
- The owner has applied for and received any required floodplain-use permit from the county flood-control district; and
- The applicant has applied for and received an aquifer-protection permit from ADEQ to recharge reclaimed water.

The underground storage facility permit includes provisions that prescribe the design capacity of the groundwater recharge project, the maximum annual amount of reclaimed water that may be stored, and monitoring requirements. The permit holder may be required to monitor groundwater-recharge project operations and the effect of storing reclaimed water on land and other water users in the project's impact area.

To store reclaimed water in an underground storage facility, a person must obtain a water-storage permit from ADWR. This permit allows the permit holder to store a specific amount of reclaimed water at a specific underground-storage facility. When issuing a water-storage permit, ADWR must determine whether the applicant has the legal right to use the reclaimed water for groundwater recharge, ensure that the reclaimed water will be stored at a permitted underground storage facility, and determine that the applicant has applied for and received an aquifer-protection permit to store the reclaimed water.

Before recovering reclaimed water that has been stored underground, the person seeking to recover the water must apply to ADWR for a recovery-well permit. If this permit is for a new well, ADWR must determine that recovering the stored water will not unreasonably increase damage to surrounding land or other water users (because of an increased concentration of wells). If the permit is for an existing well, then the applicant must demonstrate that he or she has a right to use it. A recovery-well permit includes provisions that specify the well's maximum pumping capacity.

5.3.2   Texas

Even though Texas does not have specific regulations for indirect potable reuse, any waste discharged to state waters must meet the criteria in the Texas Surface Water Quality Standards. The state has a mechanism for allowing indirect potable reuse to augment potable water resources via its alternative-system review under Texas Administration Code Section 210 (Subchapter D), which operates by a site-specific planning, piloting, monitoring, and implementation process.

In the absence of specific water quality data for microconstituents, a more general water quality evaluation approach has been developed (Allan Plummer Associates, 2004). This approach is based on the use of percent wastewater content (percent blend) and detention time as a measure of potential exposure to contaminants that may have human health effects. The use of these indicators is based on the assumption that natural degradation and dilution are important factors in reducing the quantities of potentially harmful contaminants in the water supply. Based on experience from existing planned reuse projects, a percent-blend limit of approximately 30% (average) combined with a 1-year minimum detention time (average) have been used as guidance for determining the maximum quantities of wastewater effluent that can be used to augment the supply. However, there is latitude for some variation from these criteria, particularly with the use of multiple barriers and implementation of appropriate advanced wastewater or water treatment.

One project that pursued surface augmentation is Wichita Falls, Texas. The City conducted pilot studies on membrane technology and ultimately determined that a microfiltration, ultrafiltration, reverse osmosis, and UV treatment process was acceptable for augmenting the city's water-supply lake. Water withdrawn for drinking receives more treatment (ultrafiltration to remove TDS) before distribution. This project was approved by the Texas Commission on Environmental Quality.

5.3.3 *U.S. Environmental Protection Agency Guidelines*

There are no federal regulations directly governing water reuse. At this time, U.S. water reuse practices are regulated on a state-by-state basis. The U.S. Environmental Protection Agency has published guidelines with suggested wastewater treatment processes, reclaimed water quality, monitoring, and setback distances for indirect potable reuse projects (Table 3.10) (U.S. EPA, 2004). The guidelines are not intended to be definitive water reclamation and reuse criteria or regulations, but suggested guidance for water reuse opportunities, particularly in states without their own criteria or guidelines.

6.0 RECOMMENDATIONS

This chapter provided an overview of the various health and regulatory challenges that indirect potable reuse project planners and designers face. Project proponents should keep in mind that the regulatory approach for using reclaimed water to augment potable supplies should bridge wastewater and drinking water standards with public concerns.

Before initiating a reuse project, planners and designers may want to extensively characterize the quality of the raw wastewater, reclaimed water, and any diluent water to determine the appropriate constituents to monitor continuously for treatment process reliability or SAT performance. Background monitoring of the receiving groundwater or surface water also is recommended.

Indirect potable reuse sponsors may wish to assess the potential health effects or relative risks of proposed or existing projects. There are methods available to determine possible public health effects, establish regulatory requirements, develop appropriate treatment approaches, and forge public support for water reclamation projects. These methods and strategies have various strengths and weaknesses in terms of practical application for indirect potable reuse projects.

Treatment process reliability and the use of multiple barriers will be fundamental parts of the regulatory approach. Reclaimed water is inherently suspect as a source of water supply because untreated or insufficiently treated wastewater could contain harmful contaminants (e.g., pathogens, heavy metals, and unknown organic compounds). So, regulatory controls should focus on reliable

TABLE 3.10 U.S. EPA-suggested guidelines for potable reuse projects (U.S. EPA, 2004).

Type of reuse	Treatment	Reclaimed water quality[a]	Reclaimed water monitoring	Setback distances	Comments
Groundwater recharge via surface spreading	Secondary disinfection May also need filtration or advanced wastewater treatment	Secondary disinfection Meet drinking water standards after percolation through vadose zone	Includes but not limited to: pH—daily Coliform—daily Cl$_2$ residual—continuous Drinking water standards—quarterly Other[b]—depends on constituent BOD—weekly Turbidity—continuous	150 m (500 ft) to extraction wells. May vary depending on treatment provided and site-specific conditions	The depth to groundwater should be at least 2 m (6 ft) at the maximum groundwater mounding point. The reclaimed water should be retained underground for at least 6 months before withdrawal. Recommended treatment is site specific and depends on such factors as type of soil, percolation rate, thickness of vadose zone, native groundwater quality, and dilution. Monitoring wells are necessary to detect the recharge operation's influence on groundwater. The reclaimed water should not contain measurable levels of viable pathogens after percolation through the vadose zone.[c]
Groundwater recharge via injection	Secondary Filtration Disinfection Advanced wastewater treatment	Includes but not limited to: pH = 6.5–8.5 $\leq$ 2 NTU[d] No detectable total coliform[e] 1 mg/L Cl$_2$ residual minimum[f] $\leq$ 3 mg/L TOC $\leq$ 2 mg/L TOX Meet drinking water standards	Includes but not limited to: pH—daily Turbidity—continuous Coliform—daily Cl$_2$ residual—continuous Drinking water standards—quarterly Other[b]—depends on constituent	600 m (2000 ft) to extraction wells. May vary depending site-specific conditions.	The reclaimed water should be retained underground for at least 9 months before withdrawal. Monitoring wells are necessary to detect the recharge operation's influence on groundwater. Recommended quality limits should be met at the injection point. The reclaimed water should not contain measurable levels of viable pathogens.[c] A higher chlorine residual and/or a longer contact time may be needed to ensure virus and protozoa inactivation. Refer to Sections 2.5, 2.4, and 3.43 of the U.S. EPA guidelines (2004).

TABLE 3.10 U.S. EPA-suggested guidelines for potable reuse projects (U.S. EPA, 2004) (continued).

Type of reuse	Treatment	Reclaimed water quality[a]	Reclaimed water monitoring	Setback distances	Comments
Augmentation of surface supplies	Secondary Filtration Disinfection Advanced wastewater treatment	Includes but not limited to: pH = 6.5–8.5 ≤ 2 NTU[d] No detectable total coliform[e] 1 mg/L Cl_2 residual minimum[f] ≤ 3 mg/L TOC Meet drinking water standards	Includes but not limited to: pH—daily Turbidity—continuous Coliform—daily Cl_2 residual—continuous Drinking water standards—quarterly Other[b]—depends on constituent	Site specific	Recommended level of treatment depends on such factors as receiving water quality, time and distances to withdrawal, dilution, and subsequent treatment before distribution for potable reuse. The reclaimed water should not contain measurable levels of viable pathogens.[c] A higher chlorine residual and/or a longer contact time may be needed to ensure virus and protozoa inactivation. Refer to Sections 2.6 and 3.43 of the guidelines

[a]Unless otherwise noted, recommended quality limits apply to the reclaimed water at the point of discharge from the treatment facility.

[b]Monitoring should include inorganic and organic compounds (or other classes of compounds) that are known or suspected to be toxic, carcinogenic, teratogenic, or mutagenic and are not included in the drinking water standards.

[c]It is advisable to fully characterize the reclaimed water's microbiological quality before implementing a reuse program.

[d]The recommended turbidity limit should be met before disinfection. The average turbidity should be based on a 24-hour period. Turbidity should not exceed 5 NTU at any time. If TSS is used in lieu of turbidity, it should not exceed 5 mg/L.

[e]Unless otherwise noted, recommended coliform limits are median values determined from the bacteriological results of the last 7 days for which analyses have been completed. Either the membrane-filter or fermentation-tube technique may be used. The number of fecal coliform organisms should not exceed 14/100 mL in any sample.

[f]Total chlorine residual should be met after a minimum contact time of 30 minutes.

BOD = biochemical oxygen demand
CL_2 = chlorine
NTU = Nephelometric turbidity unit
TOC = total organic carbon
TOX = total organic halogens

systems for removing these contaminants (see Chapter 4 for treatment technology and Chapter 5 for system reliability).

From a regulatory perspective, the level of treatment that should be provided depends on a number of factors:

- The type of project (typically, more treatment is required for direct-injection or surface-augmentation projects than surface-spreading projects);
- Health concerns;
- Water quality concerns;
- State regulations or requirements related to indirect potable reuse, groundwater regulation, water quality protection, and antidegradation policies; and
- Permit requirements.

Historically, the primary focus of sanitary engineering and water reclamation has been removing pathogens, so disinfection technologies are well developed, especially for bacteria. Viruses and protozoa are more resistant to disinfection, so an important aspect of water reclamation is the adequacy and reliability of treatment processes to remove, destroy, or inactivate such organisms. The reliability of disinfection systems and timely response to system failure are critical to prevent waterborne disease. So, to control microorganisms, the focus should be on treatment and disinfection performance and reliability; SAT (if applicable); and other barriers that could minimize risk (e.g., dilution with other sources of water, setback distances between spreading or injection sites and potable supply wells, and specified residence time to allow for pathogen dieoff).

Organics are another concern when using reclaimed water to augment potable supplies. While the list of organic compounds regulated in drinking water now includes numerous pesticides, industrial solvents, and disinfection byproducts, most compounds in reclaimed water are unregulated and unknown. Less than 20% of TOC in reclaimed water has been characterized. Because reclaimed water's organic component may include potentially harmful compounds or may be converted into disinfection byproducts, a cautious regulatory approach is warranted.

To control inorganic and organic chemical constituents (e.g., microconstituents), emphasis should be placed on those that can be monitored to evaluate the reliability of reclamation and reuse operations. In this sense, reliability is an assessment of a facility, or process performance, and the ability to meet water quality goals. For organics, these goals typically are expressed as a list of specific chemical limits and indicators or surrogate measures of treatment efficiency (e.g., TOC) that must be met. Another important element is implementing effective industrial pretreatment and source-control programs to restrict the release of chemicals to reclaimed water systems where controls are feasible, and public outreach programs for materials that cannot be regulated easily.

Although preliminary research indicates that human health effects due to microconstituents are unlikely at the low levels detected in drinking water, project planners and designers should be cognizant of public perception issues and incorporate a communications plan as part of any public outreach effort (see Chapter 6).

A multiple-barrier system involving demonstrated treatment technologies is essential to ensure that the reclaimed water used to augment drinking water supplies is as safe and reliable as other sources. Such barriers include dilution with other sources of water, setback distances, and a specified residence time underground or in surface water systems to allow for more inactivation or treatment, as well as time to respond if the affected potable supply becomes contaminated. Existing and draft state regulations and U.S. EPA guidelines provide insight on applicable options.

Monitoring programs for indirect potable reuse projects must be adequate to verify treatment process performance, detect potentially harmful contaminants, and collect information ensuring that the project meets applicable regulations. The results should be used to optimize treatment and implement pretreatment programs.

In general, if an indirect potable reuse project requires an NPDES permit, project planners and designers should be aware that the permitting process and permit requirements and liabilities may be more stringent than those for a non-NPDES permit, depending on the state issuing the permit. Project proponents should be aware that multiple permits may be required, sometimes with conflicting outcomes.

Project planners and designers will need to be cognizant of local water rights laws and procedures if the reclaimed water to be used is removed from a surface water. They also need to know who has the rights to the reclaimed water if it is applied to a drinking water source.

Few states have adopted regulations for indirect potable reuse projects. Information on existing or drafted regulations are presented here, along with approaches used by states in the absence of regulations. The development of more state regulations or national standards of practice may advance the implementation of indirect potable reuse projects, because project implementation would be aided by knowledge of the specific requirements that must be met. Public acceptance of using reclaimed water to augment potable water supplies may depend, in part, on having regulations in place. The public expects stringent regulatory standards that will prevent unacceptable health risks. In general, regulations should be a combination of treatment and quality requirements, with provisions for reliability, operations plans, treatment optimization, monitoring, and the preparation of an engineering report or plan providing detailed project information that can be used for regulatory approval.

7.0 REFERENCES

Allan Plummer Associates, Inc. (2004) *White River Municipal Water District Water Reuse and Conservation Study: Augmentation of Water Supply Using Reclaimed Water*; Final Report; Inc.: Fort Worth, Texas; October.

Arizona Department of Water Resources (ADWR) (1995) *Environmental Quality Act*; Sec. 49-241 of the Arizona Revised Statutes; Arizona Department of Water Resources: Phoenix, Arizona.

Asano, T.; Sakaji, R. H. (1990) Virus Risk Analysis in Wastewater Reclamation and Reuse. In *Chemical Water and Wastewater Treatment*; Hahn, H. H., Klute, R., eds.; Springer-Verlag: Berlin, Germany, 483.

Bitton, G. (1980) *Introduction to Environmental Virology*; Wiley & Sons: New York, New York.

Bouchard, D. C.; Willimans, M. K.; Surampalli, R. Y. (1992) Nitrate Contamination of Groundwater Sources and Potential Health Effects. *J. Am. Water Works Assoc.*, **84** (9), 85.

Bryan, F. L. (1974) Diseases Transmitted by Foods Contaminated by Wastewater. In *Wastewater Use in the Production of Food and Fiber—Proceedings*; Oklahoma City, Oklahoma, March 5–7; EPA-660/2-74-041, U.S. Environmental Protection Agency: Washington, D.C., 16.

Cabelli, V. J. (1983) Public Health and Water Quality Significance of Viral Diseases Transmitted by Drinking Water and Recreational Water. *Water Sci. Technol.*, **15** (5), 1–15..

California Department of Public Health (CDPH) (2001) California Health Laws Related to Recycled Water. http://www.dhs.ca.gov/ps/ddwem/waterrecycling/PDFs/purplebookupdate6-01.PDF (accessed March 2008).

California Department of Public Health (CDPH) (2007) California Groundwater Recharge Reuse Draft Regulation. http://www.dhs.ca.gov/ps/ddwem/waterrecycling/PDFs/rechargeregulationsdraft-01-04-2007.pdf (accessed March 2008).

CH2M Hill (1993) *Tampa Water Resource Recovery Project Summary Report*; Denver, Colorado.

Christensen, F. M. (1998) Pharmaceuticals in the Environment—A Human Health Risk? *Regul. Toxicol. Pharmacol.*, **28**, 212.

City of San Diego (2006) Appendix G: Science, Technology, and Regulatory Issues. In *Water Reuse Study*; San Diego, California.

Cohrssen, J. J.; Covello, V. T. (1989) *Risk Analysis: A Guide to Principles and Methods for Analyzing Health and Environmental Risks*; Council of Environmental Quality, Executive Office of the President of the United States: Washington, D.C.

Coleman, M.; Marks, H. (1998) Topics in Dose-Response Modeling. *J. Food Prot.*, **61**, 11.

Coleman, M.; Marks, H. (2000) Mechanistic Modeling of Salmonellosis. *Quant. Microbiol.*, **2**, 227–47.

Coleman, M. E.; Marks, H.; Golden, N.; Latimer, H. (2004) Discerning Strain Effects in Microbial Dose-Response Data. *J. Toxicol. Environ. Health., Part A*, **67**, 8–10.

Cooper, R. C. (1975) Wastewater Contaminants and Their Effect on Public Health. In *A "State-of-the-Art" Review of Health Aspects of Wastewater Reclamation for Groundwater Recharge*; State of California Department of Water Resources: Sacramento, California, 39.

Cooper, R. C.; Oliveri, A. W.; Danielson, R. E.; Badger, P. G.; Spear, R. C.; Selvin, S. (1986) *Evaluation of Military Field-Water Quality, Volume 5: Infectious Organisms of Military Concern Associated With Consumption: Assessment of Health Risks and Recommendations for Establishing Related Standards*; Report No. UCRL-21008; Environmental Science Division, Lawrence Livermore National Laboratory, University of California: Livermore, California.

Craun, G. F., Ed. (1986a) Recent Statistics of Waterborne Disease Outbreaks (1981–1983). In *Waterborne Diseases in the United States*; CRC Press: Boca Raton, Florida, 43.

Craun, G. F., Ed. (1986b) Statistics of Waterborne Outbreaks in the U.S. (1920–1980). In *Waterborne Diseases in the United States*; CRC Press: Boca Raton, Florida, 73.

Craun, G. F. (1991) Causes of Waterborne Outbreaks in the United States. *Water Sci. Technol.*, **24** (2), 17.

Crohn, D. M.; Yates, M. V. (1997) Interpreting Negative Virus Results from Highly Treated Water. *J. Environ. Sci. Eng.*, May, 423–430.

Crook, J.; MacDonald, J. A.; Trussell, R. R. (1999) Potable Use of Reclaimed Water. *J. Am. Water Works Assoc.*, **91** (8), 40.

Daughton, C.; Ternes, T. (1999) Pharmaceuticals and Personal Care Products in the Environment: Agents of Subtle Change? *Environ. Health Perspect.*, **107** (supplement 6), 907–938.

Dickenson, E. (2006) Identifying Indicators and Surrogates for Chemical Contaminant Removal during Indirect Potable Reuse. Presented at the 10th Annual WateReuse Foundation Research Conference; Phoenix, Arizona, May.

Ding, W. H.; Fujita, Y.; Aeschimann, R.; Reinhard, M. (1996) Identification of Organic Residues in Tertiary Effluents by GC/EI-MS, GC/CI-MS, and GC/TSQ-MS. *Fresenius' J. Anal. Chem.*, **354**, 48–55.

Donaldson, W. T. (1977) Trace Organics in Water. *Environ. Sci. Technol.*, **11** (4), 348.

Drewes, J. (2006) Viability of TOC as a Surrogate for Unregulated Trace Organics in Indirect Potable Reuse Applications. Presented at the 10th Annual WateReuse Foundation Research Conference; Phoenix, Arizona, May.

Drewes, J. E.; Heberer, T.; Reddersen, K. (2001) Removal of Pharmaceuticals during Conventional Wastewater Treatment, Advanced Membrane Treatment and Soil-Aquifer Treatment. *Proceedings of the 2nd International Conference on Pharmaceuticals and Endocrine Disrupting Chemicals in Water*; Minneapolis, Minnesota, Oct 9–11, National Groundwater Association: Westerville, Ohio.

Drewes, J.; Sedlak, D.; Snyder, S.; Dickenson, E. (2007) Identifying Indicators and Surrogates for Chemical Contaminant Removal during Indirect Potable Reuse. Presented to the California Department of Public Health's Groundwater Recharge Regulation Working Group, April.

Eisenberg, J. N.; Brookhart, M. A.; Rice, G.; Brown, M.; Colford, J. M. (2002) Disease Transmission Models for Public Health Decision Making: Analysis of Epidemic and Endemic Conditions Caused by Waterborne Pathogens. *Environ. Health Perspect.*, **110** (8), 783.

Eisenberg, J. N.; Seto, E. Y. W.; Oliveri, A. W.; Spear, R. C. (1996) Quantifying Water Pathogen Risk in an Epidemiologic Framework. *Risk Analysis*, **16** (4), 549.

EOA, Inc. (2000) *Groundwater Replenishment System Water Quality Evaluation—Risk Assessment*; EOA: Oakland, California; report prepared for the Orange County Water District, November.

Fankhauser, R. L.; Noel, J. S.; Monroe, S. S.; Ando, T.; Glass, R. I. (1998) Molecular Epidemiology of "Norwalk-like Viruses" in Outbreaks of Gastroenteritis in the United States. *J. Infect. Dis.*, **178** (6), 1571–1578.

Feachem, R. G.; Bradley, D. J.; Garelick, H. (1981) *Health Aspects of Excreta and Sullage Management: A State-of-the Art Review*; The World Bank: Washington, D.C.

Feachem, R. G.; Bradley, D. J.; Garelick, H.; Mara, D. D. (1983) *Sanitation and Disease—Health Aspects of Excreta and Wastewater Management*; Wiley & Sons: New York; published for The World Bank, Washington, D.C.

Florida Department of Environmental Protection (FDEP) (1999) Reuse of Reclaimed Water and Land Application. *Florida Administrative Code*, Chapter 17-610; Tallahassee, Florida.

Florida Department of Environmental Protection (FDEP) (2006) Reuse of Reclaimed Water and Land Application. *Florida Administrative Code*, Chapter 62-610; http://www.dep.state.fl.us/legal/rules/wastewater/62-610.pdf (accessed March 2008).

Fox, P.; Houston, S.; Westerhoff, P.; Drewes, J.; Nellor, M.; Yanko, W.; Baird, R.; Rincon, M.; Arnold, R.; Lansey, K.; Bassett, R.; Gerba, C.; Karpiscak, M.;

Amy, G.; Reinhard, M. (2001) *An Investigation of Soil Aquifer Treatment for Sustainable Water Reuse*; American Water Works Association Research Foundation: Denver, Colorado.

Fox, P.; Houston, S.; Westerhoff, P.; Nellor, M.; Yanko, W.; Baird, R.; Rincon, M.; Gully, J.; Carr, S.; Arnold, R.; Lansey, K.; Quanrud, D.; Ela, W.; Amy, G.; Reinhard, M., Drewes, J. (2006) *Advances in Soil–Aquifer Treatment for Sustainable Reuse*; American Water Works Association Research Foundation: Denver, Colorado.

Fuhs, G. W. (1975) A Probabilistic Model of Bathing Beach Safety. *Sci. Total Environ.*, **4**, 165.

Garcia, A.; Yanko, W.; Batzer, G.; Widmer, G. (2002) *Giardia* cysts in Tertiary-Treated Wastewater Effluents: Are they Infective? *Water Environ. Res.*, **74**, 541–544.

Gennaccaro, A. L.; McLaughlin, M. R.; Quintero-Bentancourt, W.; Huffman, D. E.; Rose, J. B. (2003) Infectious *Cryptosporidium parvum* Oocysts in Final Reclaimed Effluent. *Appl. Environ. Microbiol.*, **69** (8), 4983–4984; August.

Gerba, C. P.; Goyal, S. M. (1985) Pathogen Removal from Wastewater during Groundwater Recharge. In *Artificial Recharge of Groundwater*; Asano, T., Ed.; Butterworth Publishers: Boston, Massachusetts, 283.

Gerba, C. P.; Rose, J. B.; Hass, C. N.; Crabtree, K. D. (1996) Waterborne Rotavirus: A Risk Assessment. *Water Res.*, **30** (12), 2929.

Gharravi A. M.; Ghorbani, R.; Khazaei, M.; Motabbad, P. A.; Al Agha, M.; Ghasemi, J.; Sayadi, P. (2006) Altered Pituitary Hormone Secretion in Male Rats Exposed to Bisphenol A. *Indian J. Occup. Environ. Med.*, 10, 24–27.

Goyal, S. M.; Gerba, C. P. (1979) Comparative Adsorption of Human Enteroviruses, Simian Rotavirus, and Selected Bacteriophages to Soils. *Appl. Environ. Microbiol.*, **38**, 242.

Gruener, N. (1979) Biological Evaluation of Toxic Effects of Organic Contaminants in Concentrated Recycled Water. In *Water Reuse Symposium Proceedings—Vol. 3*, Washington, D.C., March 25–30; American Water Works Association Research Foundation: Denver, Colorado.

Haas, C. N. (1983) Estimation of Risk Due to Low Doses of Microorganisms: A Comparison of Alternative Methodologies. *Am. J. Epidemiol.*, **55**, 573.

Halling-Sørensen, B.; Nielsen, S. N.; Lanzky, P. F.; Ingerslev, F.; Lutzhoft, H. C. H.; Jorgensen S. E. (1998) Occurrence, Fate, and Effects of Pharmaceutical Substances in the Environment—A Review. *Chemosphere*, **36**, 357–394.

Hatch, M. C.; Beya, J.; Nieves, J. W.; Susser, M. (1990) Cancer Near the Three Mile Island Nuclear Plant: Radiation Emissions. *Am. J. Epidemiol.*, **132**, 397.

Heberer, T. K.; Reddersen, K.; Mechlinski, A. (2002) From Municipal Sewage to Drinking Water: Fate and Removal of Pharmaceutical Residues in the Aquatic Environment in Urban Areas. *Water Sci. Technol.*, **46** (3), 81–88.

Herwaldt, B. L.; Craun, G. F.; Stokes, S. L.; Juranek, D. D. (1992) Outbreaks of Waterborne Disease in the United States: 1989–1990. *J. Am. Water Works Assoc.*, **84**, 129.

Hurst, C. J.; Benton, W. H.; Stetler, R. E. (1989) Detecting Viruses in Water. *J. Am. Water Works Assoc.*, **91**, 71.

Idaho Department of Environmental Quality (IDEQ) (2007a) Ground Water Quality Rule; IDAPA 58.01.11. http://adm.idaho.gov/adminrules/rules/idapa58/0111.pdf (accessed March 2008).

Idaho Department of Environmental Quality (IDEQ) (2007b) Rules for the Reclamation and Reuse of Municipal and Industrial Wastewater; IDAPA 58.01.17. Posted on the Internet at http://adm.idaho.gov/adminrules/rules/idapa58/0117.pdf (accessed March 2008).

Isaacson, M.; Sayed, A. R. (1988) Health Aspects of the Use of Recycled Water in Windhoek, SWA/Namibia, 1974–1983. *S. African Med. J.*, **73**, 596–599.

James M. Montgomery, Inc. (1983) *Operation, Maintenance, and Performance Evaluation of the Potomac Estuary Wastewater Treatment Plant*; James M. Montgomery, Inc.: Alexandria, Virginia.

Keller, W. (2002) Reuse of Stormwater and Wastewater in the City of Calgary. Presentation at CCME Workshop; Calgary, Alberta, Canada, May 30–31.

Khan, S.; Roser, D. (2007) *Risk Assessment and Health Effects Studies of Indirect Potable Reuse Schemes: Final Report*; CWWT Report 2007/01; School of Civil and Environmental Engineering, University of New South Wales: Kensington, New South Wales, Australia; prepared for Local Government Association of Queensland (LGAQ) Centre for Water and Waste Technology.

Koplin, D. W.; Furlong, E. T.; Meyer, M. T.; Thurman, M.; Zaugg, S. D.; Barber, L. B.; Buxton, H. T. (2002) Pharmaceutical, Hormones, and Other Organic Wastewater Contaminants in U.S. Streams, 1999–2000: A National Reconnaissance. *Environ. Sci. Technol.*, **36** (6), 447–451.

Kramer, M. H.; Herwaldt, B. L.; Craun, G. F.; Calderon, R. L.; Juranek, D. D. (1996) Waterborne Disease: 1993 and 1994. *J. Am. Water Works Assoc.*, **88**, 66.

Lauer, W. C.; Rogers, S. E. (1996) The Demonstration of Direct Potable Water Reuse: Denver's Pioneer Project. *Proceedings of the AWWA/WEF 1996 Water Reuse Conference*; San Diego, California, Feb 25–28; American Water Works Association: Denver, Colorado; 269.

Levine, W. C.; Stephenson, W. T. (1990) Waterborne Disease Outbreaks, 1986–1988. *Morb. Mort. Wkly. Rpt.*, **39** (SS-1).

Mead, P. S.; Slutsker, L.; Dietz, V.; McCaig, L. F.; Bresee, J. S.; Shapiro, C.; Griffin, P. M.; Tauxe, R. V. (1999) Food-Related Illness and Death in the United States. *Emerg. Infect. Dis.*, **5** (5), 607–625.

Mena, K. D.; Gerba, C. P.; Haas, C. N.; Rose, J. B. (2003) Risk Assessment of Waterborne Cocksackievirus. *J. Amer. Water Works Assoc.*, **95** (7), 122.

Metcalf and Eddy, Inc. (2002) *Wastewater Engineering: Treatment, Disposal, Reuse*, 4th ed.; McGraw-Hill: New York.

Mitch, W.; Sedlak, D. L. (2003) Fate of N-nitrosodimethylamine (NDMA) Precursors during Municipal Wastewater Treatment. *Proceedings of the American Water Works Association Annual Conference*; Anaheim, California, June 15–19; American Water Works Association: Denver: Colorado.

Mofidi, A. A.; Meyer, E. A.; Wallis, P. M.; Chou, C. I.; Meyer, B. P.; Ramalingam, S.; Coffey, B. M. (2002) The Effect of UV Light on the Inactivation of *Giardia lamblia* and *Giardia muris* cysts as Determined by Animal Infectivity Assay. *Water Res.*, **36**, 2098–2108.

Moolgavkar, S. H. (1994) Biological Models of Carcinogenesis and Quantitative Cancer Risk Assessment. *Risk Anal.*, **14**, 879.

National Research Council (NRC) (1994) *Ground Water Recharge Using Waters of Impaired Quality*; National Academy Press: Washington, D.C.

National Research Council (NRC) (1998) *Issues in Potable Reuse—The Viability of Augmenting Drinking Water Supplies with Reclaimed Water*; National Academy Press: Washington, D.C.

Nellor, M. H.; Baird, R. B.; Smyth, J. (1984) *Health Effects Study*; Final Report; County Sanitation Districts of Los Angeles County: Whittier, California.

Olivieri, A. W.; Cooper, R. C.; Spear, R. C.; Selvin, S.; Danielson, R. E.; Block, D. E.; Badger, P. G. (1986) Risk Assessment of Waterborne Infectious Agents. *Proceeding of ENVIROSOFT 86: International Conference on Development and Applications of Computer Technology to Environmental Studies*; Newport Beach, California, Nov 19–21; appears in *Proceedings of the International Conference on Development and Application of Computer Techniques to Environmental Studies*; CML Publications: Ashurst, Southampton, U.K.; 601–607.

Oscar, T. (2004) Dose–Response Model for 13 Strains of *Salmonella*. *Risk Anal.*, **24** (1).

Palmateer, G. A., Dutka, B. J., Janzen, E. M., Meissner, S. M., Sakellaris, M. G. (1991) Coliphage and Bacteriophage as Indicators of Recreational Water Quality. *Water Res.*, **25** (3), 355–357.

Pereira, M. A.; Castro, B. C.; Tabor, M. W.; Khoury, M. D. (undated) Toxicological Studies of the Tampa Water Resources Project. Report supplied to the 1998

Committee to Evaluate the Viability of Augmenting Potable Water Supplies with Reclaimed Water, National Research Council, by M. Perreira, Medical College of Ohio, Toledo, Ohio

Polissar, L. (1980) The Effect of Migration on Comparison of Disease Rates in Geographic Studies in the United States. *Am. J. Epidemiol.*, **111**, 175.

Powelson, D. K.; Gerba, C. P.; Yahya, M. T.(1993) Virus Transport and Removal in Wastewater during Aquifer Recharge. *Water Res.*, **27**, 583–590.

Quintero-Betancourt, W.; Gennaccaro, A. L.; Scott, T. M.; Rose, J. B. (2003) Assessment of Methods for Detection of Infected *Cryptosporidium* Oocysts and *Giardia* Cysts in Reclaimed Effluent. *Appl. Environ. Microbiol.*, **69**, 5380–5388.

Rao, V. C.; Melnick, J. L. (1986) *Environmental Virology*; American Society of Microbiology: Washington, D.C.

Regli, S.; Rose, J. B.; Haas, C. N.; Gerba, C. P. (1991) Modeling the Risk from *Giardia* and Viruses in Drinking Water. *J. Am. Water Works Assoc.*, **83**, 76.

Reuse Committee of the Texas Water Conservation Association (2006) *Texas Water Rights and Wastewater Reuse*; Texas Water Conservation Association: Austin, Texas.

Richardson, M. L.; Bowron, J. M. (1985) The Fate of Pharmaceutical Chemicals in the Aquatic Environment. *J. Pharm. Pharmacol.* **37**, 1–12.

Rose, J. B.; Carnahan, R. P. (1992) *Pathogen Removal by Full-Scale Wastewater Treatment.* Report prepared for the Florida Department of Environmental Regulation, Tallahassee, Florida.

Rose, J. B.; Gerba, C. P. (1991) Use of Risk Assessment for Development of Microbial Standards. *Water Sci. Technol.*, **24**, 2.

Rose, J. B.; Quintero-Betancourt, W. (2002) *Monitoring for Enteric Viruses,* Giardia, Cryptosporidium, *and Indicator Organisms in the Key Colony Beach Wastewater Treatment Plant Effluent*; University of South Florida: St. Petersburg, Florida.

Rose, J. B.; Haas, C. N.; Regli, S. (1991) Risk Assessment and Control of Waterborne Giardiasis. *Am. J. Public Health*, **81**, 709.

Rose, J. B.; Huffman, D. E.; Riley, K.; Farrah, S. R.; Lukasik, J. O.; Hamann, C. L. (2001) Reduction of Enteric Microorganisms at the Upper Occoquan Sewage Authority Water Reclamation Plant. *Water Environ. Res.*, **73**, 6.

Sagik, B. P.; Moore, B. E.; Sorber, C. A. (1978) Infectious Disease Potential of Land-Application of Wastewater. *Proceedings of the International Symposium on the State of Knowledge in Land Treatment of Wastewater*; U.S. Army Corps of Engineers, Cold Reg. Research Engineering Laboratory: Hanover, New Hampshire; p. 35.

Santa Ana Regional Water Quality Control Board (SARWQCB) (2004) Resolution Amending the Water Quality Control Plan for the Santa Ana River Basin to Incorporate an Updated Total Dissolved Solids (TDS) and Nitrogen Management Plan for the Santa Ana Region; R8-2004-001; Riverside, California.

Schwab, B. W.; Hayes, E. P.; Fiori, J. M.; Mastrocco, F. J.; Roden, N. M.; Cragin, D.; Meyerhoff, R. D.; D'Acoand, V. J.; Anderson, P. D. (2005) Human Pharmaceuticals in U.S. Surface Waters: A Human Health Risk Assessment. *Regul. Toxicol. Pharmacol.*, **42**, 296.

Sheikh, B.; Cooper, R. C. (1998) *Recycled Water Food Safety Study*. Report to the Monterey County Water Resources Agency and Monterey Regional Water Pollution Control Agency.

Sheikh, B.; Cooper R. C.; Israel K. E. (1999) Hygienic Evaluation of Reclaimed Water Used to Irrigate Food Crops—A Case Study. *Water Sci. Technol.*, **40**, 261–267.

Singer, P. C. (1994) Issues and Concerns for the Control of Disinfection Byproducts. *J. Environ. Eng.*, **120** (4), 727.

Slifko, T. R. (2001) Development and Evaluation of a Quantitative Cell Culture Assay for *Cryptosporidium* Disinfection Studies. Ph.D. Dissertation, University of South Florida, College of Marine Science, St. Petersburg, Florida.

Sloss, E. M.; Geschwind, S. A.; McCaffrey, D. F.; Ritz, B. R. (1996) *Groundwater Recharge with Reclaimed Water: An Epidemiologic Assessment in Los Angeles, County, 1987-1991*; RAND: Santa Monica, California. Report prepared for the Water Replenishment District of Southern California.

Sloss E. M.; McCaffrey, D. F.; Fricker, R. D.; Geschwind, S. A.; Ritz, B. R. (1999) *Groundwater Recharge With Reclaimed Water: Birth Outcomes in Los Angeles County 1982–1993*. Report prepared for the Water Replenishment District of Southern California, by RAND, Santa Monica, California.

Snyder, S. (2005) Toxicological Relevance of EDCs and Pharmaceuticals in Water; AWWARF Project #3085 and WRF Project 04-003. Paper presented at 9th Annual WateReuse Foundation Research Conference; Denver, Colorado, May.

Snyder, S. (2006) Removal of Wastewater-Derived Contaminates during Oxidation Processes. Paper presented at the 10th Annual WateReuse Foundation Research Conference, Phoenix, Arizona, May.

Snyder, S. (2007a) Endocrine Disruptors & Pharmaceuticals: Implications for the Water Industry. Presentation at the California State Water Resources Control Board, Sacramento, California; May.

Snyder, S. (2007b) Toxicological Relevance of Endocrine-Disrupting Chemicals and Pharmaceuticals in Water. Presentation to the California Urban Water Agencies; June.

Sobsey, M.; Battigelli, D. A.; Handzel, T. R.; Schwab, K. J. (1995) *Male-Specific Coliphages as Indicators of Viral Contamination of Drinking Water*; American Water Works Association Research Foundation: Denver, Colorado.

Soller, J. A.; Olivieri, A. W.; Eisenberg, J. N. S.; Sakaji, R.; Danielson, R. (2004) *Evaluation of Microbial Risk Assessment Techniques and Applications*; Project 00-PUM-3; Water Environment Research Foundation: Alexandria, Virginia.

Soller, J. A.; Seto, E.; Olivieri, A. W. (2007) *Microbial Risk Assessment Interface Tool*; Project 04-HHE-3; Water Environment Research Foundation: Alexandria, Virginia.

Soller, J. A. (2006) Use of Microbial Risk Assessment to Inform the National Estimate of Acute Gastrointestinal Illness Attributable to Microbes in Drinking Water. *J. Water Health*, 04 (Suppl. 2), 165.

Stan, H. J.; Heberer, T. (1997) Pharmaceuticals in the Aquatic Environment. In *Dossier Water Analysis*; Suter, M. J. F., Coord.; *Analysis*, 25, M20–23.

State of Arizona (1991) *Regulations for the Reuse of Wastewater*. Arizona Administrative Code, Chapter 9, Article 7; Arizona Department of Environmental Quality: Phoenix, Arizona.

State of California (1987) *Groundwater Recharge with Reclaimed Water*; Robeck, G., ed.; Department of Water Resources: Sacramento, California; Scientific Advisory Panel report prepared for State of California, State Water Resources Control Board, Department of Water Resources, and California Department of Health Services.

Straub, T. M.; Pepper, I. L.; Gerba, C. P. (1993) Hazards from Pathogenic Microorganisms in Land-Disposed Sewage Sludge. *Rev. Environ. Contam. Toxicol.*, **132**, 55–93.

Tanaka, H.; Asano, T.; Schroeder, E.; Tchobanoglous, G. (1998) Estimating the Reliability of Wastewater Reclamation and Reuse Using Enteric Virus Monitoring Data. *Water Environ. Res.*, **70**, 39–51.

Templer, O. W. (1991) Water Rights Issues—Texas Water Rights Law: East Meets West. *Journal of Contemporary Water Research and Education*, **85** (13), 13–18.

Ternes, T. (1998) Occurrence of Drugs in German Sewage Treatment Plants and Rivers. *Water Res.*, **32** (11), 3245–3260.

Teunis, P. F.; Havelaar, A. H. (2000) The Beta Poisson Dose-Response Model is Not a Single-Hit Model. *Risk Anal.*, **20** (4), 513–520.

Teunis, P. F. M.; van der Heijden, O. G.; van der Giessen, J. W. B.; Havelaar, A. H. (1996) *The Dose–Response Relation in Human Volunteers for Gastro-intestinal Pathogens*; RIVM Report 284550002; Rijksinstituut voor Volksgezondheid en Milieu, Bilthoven, The Netherlands.

Teunis, P.; Takumi, K.; Shinagawa, K. (2004) Dose Response for Infection by *Escherichia coli* O157:H7 from Outbreak Data. *Risk Anal.*, **24** (2), 401–407.

U.S. Environmental Protection Agency (1974) *Design Criteria for Mechanical, Electric, and Fluid System and Component Reliability*, MCD-05; EPA-430/99-74-001; Washington, D.C.

U.S. Environmental Protection Agency (1976) *National Interim Primary Drinking Water Standards*; EPA-570/9-76-003; Washington, D.C.

U.S. Environmental Protection Agency (2004) *Guidelines for Water Reuse*; EPA-625/R-04-108; Washington, D.C.

U.S. Environmental Protection Agency (2006a) National Primary Drinking Water Regulations: Ground Water Rule. *Fed. Regist.*, **71** (216), 65574–65660.

U.S. Environmental Protection Agency (2006b) *National Recommended Water Quality Criteria*; 4304T; D.C.

U.S. Environmental Protection Agency (2006c) National Primary Drinking Water Regulations: Long Term 2 Enhanced Surface Water Treatment Rule. *Code of Federal Regulations*, Parts 9, 141, and 142, Vol. 40; *Fed. Regist.*, **71** (654); Jan. 5.

Walker-Coleman, L. (2002) Protozoan Pathogen Monitoring Results for Florida's Reuse Systems. *Proceedings of Symposium XVII*; Orlando, Florida, Sept. 8–10; WateReuse Association: Alexandria, Virginia.

Washington Department of Health; Washington Department of Ecology (1997) Water Reclamation and Reuse Standards; Publication #97-23. Available on the Internet at http://www.ecy.wa.gov/programs/ (accessed March 2008).

Wellings, F. M. (1980) Presentation to the Florida Environmental Regulation Commission; Orlando, Florida, May 14.

Western Consortium for Public Health (1996) *Total Resource Recovery Project Final Report*. City of San Diego, California; Western Consortium for Public Health in association with EOA, Inc.: Oakland, California.

World Health Organization (WTO) (1999) *Health-Based Monitoring of Recreational Waters: The Feasibility of a New Approach (The 'Annapolis Protocol')*; WHO/SDE/WSH/99.1; Geneva.

Yanko, W. (1993) Analysis of 10 Years of Virus Monitoring Data from Los Angeles County Treatment Plants Meeting California Wastewater Reclamation Criteria. *Water Environ. Res.*, **65** (3), 221–226.

Yates, M. V. (1994) Monitoring Concerns and Procedures for Human Health Effects. In *Wastewater Reuse for Golf Course Irrigation*; Lewis Publishers: Chelsea, Michigan; p. 143.

York, D. W. (2002) Pathogens in Reclaimed Water: The Florida Experience. *Proceedings of Water Sources 2002*; Las Vegas, Nevada, Jan 27–30; American Water

Works Association: Denver, Colorado; Water Environment Federation: Alexandria, Virginia.

8.0   SUGGESTED READINGS

Crook, J. T.; Asano, T.; Nellor, M. H. (1990) Groundwater Recharge with Reclaimed Water in California. *Water Environ. Technol.*, **28** (2), 42–49.

State of California (1976) *Report of the Consulting Panel on Health Aspects of Wastewater Reclamation for Groundwater Recharge*; Department of Water Resources: Sacramento, California; report prepared for the California State Water Resources Control Board, Department of Water Resources, and California Department of Health.

Chapter 4

Treatment Technology

1.0   INTRODUCTION

This chapter reviews advanced treatment technologies for use in meeting indirect potable reuse water requirements. Preliminary, primary, and secondary treatment will not be addressed; it is assumed that these unit processes will be provided ahead of advanced treatment. The treatment processes necessary will depend on the specific method used to augment a potable water resource. The treatment requirements that will be discussed are for *indirect potable reuse*—the planned discharge of reclaimed water to replenish or augment a surface-water or groundwater source of potable water.

Reclaimed water-quality requirements are not well defined. The level of treatment required depends on many factors (in addition to the regulatory factors discussed in Chapter 3):

- The source and quality of reclaimed water,
- Site-specific conditions
- Groundwater quality,
- Surface water quality,
- The reuse application, and
- The dilution factor.

The treatment technologies may be required to remove or reduce the following from wastewater:

- Pathogens,
- Nutrients,
- Trace organics,
- Trace metals,
- Total dissolved solids (TDS), and
- Microconstituents.

At a minimum, the selected treatment processes should be capable of producing effluent that meets the quality of the existing drinking water supply (McEwen and Richardson, 1996). Because there are no specific standards for indirect potable reuse, the concept of "multiple barriers" (McEwen and Richardson, 1996) should be used to develop treatment trains that will provide the necessary level of safety and reliability for the public. It consists of using multiple, redundant treatment processes to reduce, remove, or inactivate contaminants of concern. If a contaminant escapes one barrier, others downstream are available to remove it. Selecting multiple-barrier treatment processes will depend on several factors, including

- Proven performance;
- Economics (e.g., capital cost, operations and maintenance cost, lifetime cost);

- The system's redundancy, flexibility, and operating complexity;
- Land availability;
- Regulatory requirements; and
- Efficiency.

The combination of treatment processes typically used to provide the treatment required for augmenting potable water supplies consists of the following:

- Preliminary and primary treatment,
- Secondary treatment,
- Advanced treatment,
- Natural storage and/or treatment system,
- Disinfection, and
- Water reclamation solids management.

The treatment processes required also will depend on the level of industrial pretreatment provided upstream of the water reclamation facility. This is an important consideration given that industrial discharges may contain priority pollutants and other toxics that pose a reasonable risk of interfering with or passing through municipal wastewater treatment. So, a detailed review of potential industrial discharges and pretreatment program adequacy should be made before selecting a treatment process for the multiple-barrier approach.

In addition to liquid treatment processes, a water reclamation system's final design also must include a method for handling residuals. The total residuals to be handled will increase incrementally with each successive unit process added to the treatment train. The ultimate disposal of residuals is beyond the scope of this chapter, but still important when developing a complete treatment system.

The rest of this chapter will provide a brief description of the purpose and objectives of advanced treatment, disinfection, and natural treatment systems. However, the main focus will be on advanced treatment processes because these are incremental processes beyond secondary wastewater treatment to provide the added measure of protection for indirect potable reuse. Design manuals and procedures should be consulted for more detailed information on specific treatment process applications.

2.0   ADVANCED TREATMENT

Advanced treatment processes remove or further reduce constituents in reclaimed water that were partially reduced via conventional secondary treatment. When handling reclaimed water to augment potable water supplies, an advanced treatment process may be designed to reduce total dissolved solids and/or remove viruses and other pathogens, nutrients (e.g., nitrogen and phosphorus), trace metals, organics, or microconstituents.

Such processes include

- Biological nutrient removal (e.g., activated sludge, biologically active filter);
- Chemically enhanced nutrient removal;
- Coagulation/flocculation;
- Membrane filtration (e.g., microfiltration, ultrafiltration, nanofiltration, and reverse osmosis);
- Membrane bioreactor (low-pressure membranes);
- Ion exchange;
- Electrodialysis and electrodialysis reversal (current-driven membrane separations);
- Chemical oxidation (ozone);
- Advanced oxidation processes (AOPs) (e.g., ozone/UV irradiation, ozone/hydrogen peroxide, ozone/UV/hydrogen peroxide, UV/hydrogen peroxide, and hydrogen peroxide/chlorine);
- UV radiation (using medium-pressure or low-pressure, high-output lamps);
- Activated carbon adsorption;
- Air stripping; and
- Natural treatment systems (e.g., constructed wetlands, soil–aquifer treatment, and riverbank filtration).

Table 4.1 summarizes these processes and the principal contaminants they remove. Combinations of these processes will provide multiple barriers to contaminants.

TABLE 4.1   Removal mechanisms of certain pollutants via soil–aquifer treatment and via conventional treatment.

Pollutant	Conventional treatment removal mechanism	Soil–aquifer treatment removal mechanism
Viruses/pathogens	• Flocculation • Filtration • Disinfection	• Precipitation • Flocculation • Filtration
Nitrogen	• Biological nitrification/denitrification	• Aerobic and anaerobic biological degradation
Organic compounds • Volatile organics • Semi- and non-volatile organics	• Air stripping • Granular activated carbon adsorption	• Volatilization • Adsorption
Inorganics/metals	• Coagulation • Precipitation/flocculation • Filtration • Tertiary sedimentation	• Precipitation • Adsorption • Filtration

2.1   Biological and Chemical Nutrient Removal

Removing nutrients from wastewater effluents has become a primary concern from an environmental protection standpoint. Most countries have eutrophication-related problems. High concentrations of nutrients (e.g., phosphorus and nitrogen) lead to eutrophication and other unwanted events, so phosphorus and nitrogen discharges to receiving waters need to be regulated and, in most cases, minimized. Most U.S. plants must demonstrate a low level of effluent toxicity by testing the effluent via aquatic species (e.g., daphnia and fathead minnows). One contaminant of concern is un-ionized ammonia. Reducing ammonia in effluent implies that many facilities have been required to convert high-rate activated-sludge plants to nitrifying plants. It has often been economical to incorporate denitrification in these plants due to oxygen and alkalinity recovery, thereby resulting in nitrogen control.

Maintaining the quality of potable water supplies is also a concern for areas with limited water resources. In the southwestern United States, for example, nitrate-contaminated groundwater is a concern, so many plants must remove nitrogen if the effluent is for reuse purposes (Bratby, 1996).

Nitrogen may be removed via chemical or biological means. Chemical methods (e.g., ammonia stripping or ion exchange) have not become common because of their high operating costs, quantities of residuals produced, and lesser effluent quality compared to biological nutrient-removal systems. Recently, biologically active filters or biological aerated filtration have been used in water reclamation plants to remove more nitrogen (e.g., Denver Water's Water Reclamation Plant).

Biological nitrogen removal is a two-step process. First, ammonia is oxidized to nitrate via nitrification. Second, nitrate is made to serve as an electron acceptor (in place of oxygen) for biological respiration so it can be reduced to molecular nitrogen via denitrification.

Ammonia principally is oxidized to nitrate by two groups of autotrophic bacteria: *Nitrosomonas* and *Nitrobacter*. *Nitrosomonas* oxidizes ammonia to nitrite, and *Nitrobacter* oxidizes nitrite to nitrate. The oxygen required for ammonia oxidation is approximately 4.57 g/g of nitrogen oxidized, with 3.2 g/g of nitrogen used to oxidize ammonia to nitrite and 1.1 g/g of nitrogen used to oxidize nitrite to nitrate. Alkalinity is also consumed during nitrification (7.2 ppm alk/ppm of ammonia-nitrogen oxidized).

During denitrification, various bacteria groups use the oxygen associated with nitrate for metabolic processes, converting the nitrate to nitrogen gas. Most denitrifying bacteria are heterotrophic and use the organic carbon in water as a carbon source. Typically, these bacteria constitute most of the total bacteria mass. Denitrification occurs only when the dissolved oxygen content is low (anoxic conditions) and nitrate is present, because most of the bacteria are facultative and prefer oxygen, if present, as an oxidizing agent. Denitrification reduces nitrate to nitrite to nitric oxide to nitrous oxide to nitrogen as follows:

$$NO_3 \rightarrow NO_2 \rightarrow NO \rightarrow N_2O \rightarrow N_2 \qquad (4.1)$$

Denitrification occurs under anoxic conditions and requires an organic or inorganic electron donor. The measured bulk liquid dissolved-oxygen concentration does not represent the actual dissolved-oxygen concentration in a microcosm (e.g., an activated-sludge floc). Although it is an anoxic reaction, denitrification still occurs under aerated conditions with bulk liquid dissolved-oxygen concentrations of up to approximately 0.2 mg/L.

Including denitrification in a nitrifying activated-sludge system may reduce the overall oxygen requirement and recapture some of the alkalinity used in nitrification, depending on the process configuration. Approximately 3.55 g of alkalinity are recovered for every gram of nitrate denitrified. If wastewater is used as the carbon source, denitrification also reduces the oxygen demand of the wastewater's carbonaceous material by 2.86 g for every gram of nitrate denitrified. If methanol is used, "oxygen demand reduction" does not occur. Aeration-energy reductions of 20 to 30% are typical when denitrification is incorporated in a nitrifying activated-sludge plant. Approximately one-half of the alkalinity consumed during nitrification is recovered via denitrification.

Current experience and the design approach followed in the Mid-Atlantic (Chesapeake Bay) indicates that biological nutrient removal systems can readily achieve less than 8 mg/L of nitrogen via nitrification–denitrification processes—less than 5 mg/L with supplemental carbon addition. Denitrification filters can be added to further reduce nitrate and ensure total nitrogen levels of less than 3 mg/L; however, some facilities are achieving or approaching 3 mg/L total nitrogen using only activated sludge.

Chemical phosphorus removal is widely used in North America, South Africa, and Northern Europe. Metal coagulants (e.g., ferric chloride, ferric sulfate, aluminum sulfate, sodium aluminate, and polyaluminum chloride) typically are used. Lime also has been used but has lost favor because it produces large quantities of sludge and alkaline effluent.

Coagulant addition at a tertiary stage is often the preferred method for low final effluent phosphorus concentrations. Because biosolids typically contain at least 2 to 3% phosphorus by mass, tertiary filtration, low-pressure membranes, or dissolved air flotation is often required to reduce effluent solids to achieve effluent total phosphorus levels that are significantly less than 0.5 mg/L. Coagulant addition and tertiary filtration are often provided as a polishing step after biological phosphorus removal, especially when effluent phosphorus levels significantly less than 1 mg/L must be maintained. In tertiary phosphorous removal, considerable savings in chemical costs are possible because coagulants are only required for the residual phosphorus.

During biological phosphorus removal, an anaerobic zone is included in the activated-sludge reactor to select particular strains of bacteria (e.g., *Acinetobacter*). These bacteria use short-chain volatile fatty acids (e.g., acetates), which constitute a portion of the soluble chemical oxygen demand entering the plant, and store them within the bacterial cell as polyhydroxybutyrate (PHB). The energy for this uptake is gained via hydrolysis of polyphosphates stored in the cells.

During this process, hydrolyzed polyphosphates are lost from the cells into the surrounding mixed liquor as orthophosphates. This orthophosphate release into anaerobic-zone bulk water can be three to five times the influent phosphorus concentration. When polyphosphorus organisms travel to the reactor's aerobic zone, the stored short-chain volatile fatty acid substrate is depleted, and phosphorus is taken up and stored as polyphosphates, ready for the next cycle of short-chain volatile fatty acids uptake and PHB storage in the anaerobic zone. Phosphorus is removed from the system by wasting a portion of the phosphorus-rich cells from the anaerobic zone along with the waste activated sludge.

2.2 Coagulation/Flocculation and Solid–Liquid Separation

Coagulation and solid–liquid separation is a chemical/physical process in which a chemical (coagulant) is added to water to produce flocs of suspended particulate and precipitate other contaminants (e.g., heavy metals, phosphorus, and organics). The process typically involves the following three steps:

- Rapid mixing to efficiently disperse chemicals in water;
- Slow mixing to promote formation of large settleable flocs (flocculation); and
- Separation of flocs containing suspended solids, colloids, or precipitates from bulk water in sedimentation, and separation of precipitates via sedimentation, granular media filtration, dissolved-air flotation tanks, media filtration, low-pressure membranes (microfiltration/ultrafiltration), or a combination of these technologies.

Chemicals used for coagulants include alum, ferric chloride, lime, polymers, PACl, polyelectrolytes, polymer flocculants, and various prehydrolyzed aluminum or iron salts. Each chemical has unique properties and applications (Bratby, 1980).

High-pH lime clarification has been practiced at a number of potable reuse projects (e.g., Water Factory 21 and West Basin Water Recycling Plant in Southern California and Denver's Potable Water Reuse Demonstration Plant). During lime clarification, calcium hydroxide (lime) is added to water to raise the pH above 11 to form an insoluble precipitate. The precipitate is flocculated and then removed via settling. Lime clarification may be followed by recarbonation. Carbon dioxide is added to provide calcium carbonate equilibrium and lower the pH. In addition to metals, phosphate, and organics removal, high-pH lime clarification has been shown to be a barrier to viruses, bacteria, and protozoa. Because of the excess sludge production, this process has been replaced by low-pressure membranes at Water Factory 21 (now called the Groundwater Replenishment District) and West Basin.

Enhanced coagulation is the process in which coagulant is injected in higher doses than required for particulate removal (expressed by total suspended solids or turbidity). Enhanced coagulation allows for the optimized removal of natural organic matter, as measured by total organic carbon (TOC), because of the organics sorption by the created flocs, thereby reducing disinfection byproducts. Enhanced coagulation still maintains the objective of suspended particulate removal (measured as turbidity) but also is designed to enhance TOC removal.

Natural and effluent organic matter (NOM, EfOM) is removed via adsorption onto aluminum hydroxide or ferric hydroxide flocs and via the formation of insoluble complexes between aluminum or iron species and humic or fulvic acids. Adsorption onto aluminum hydroxide flocs typically occurs at higher coagulant doses and higher pH. Complex formation between metal coagulant species and NOM/EfOM compounds typically is more dominant at lower coagulant doses and lower pH. Optimizing enhanced coagulation requires optimizing coagulant type, dose, and pH to reduce turbidity and remove organic matter.

2.3   Microfiltration and Ultrafiltration

Microfiltration and ultrafiltration membranes can be either vacuum (immersed, submerged) or pressurized type, configured as hollow fibers or flat sheets. Microfiltration is a pressure-driven or vacuum membrane filtration process that targets removal of all particulate within the 0.1 to 10 μm range. Microfiltration is an effective barrier to bacteria, *Giardia*, and *Cryptosporidium*. The process can, therefore, reduce downstream disinfection requirements, thereby reducing the formation of disinfection byproducts. Although the process is not designed for complete virus removal, California has accepted microfiltration for meeting the Surface Water Treatment Rule and granted 0.5-log or greater credits for virus removal and up to 4.0 logs on pathogen removal (e.g., *Giardia* cysts and *Cryptosporidium* oocysts), depending on the membrane brand and model (4.0 logs removal = 99.99% removal, 3.0 logs = 99.9%, etc.). The U.S. Environmental Protection Agency *Membrane Filtration Guidance Manual* can be used for guidance, but each state may have its own set of guidelines to follow and should be referenced.

Ultrafiltration operates in a smaller filtration range [about 0.01 to 0.1 μm or approximately 10 000 to 150 000 Dalton molecular weight cutoff (MWCO)] and typically is used to remove oils, colloids, and high-molecular-weight organics. In

California, ultrafiltration membranes are credited with up to 2.0- to 4.0-log virus removal and up to 4.0-log pathogen removal (e.g., *Giardia* cysts and *Cryptosporidium* oocysts), depending on membrane characteristics. The California Department of Health Services established test protocol on a microfiltration and ultrafiltration membrane-certification process. Microfiltration and ultrafiltration membranes differ from reverse osmosis and nanofiltration membranes not only in the particle size removed but also in operating pressure and residual characteristics. Microfiltration and ultrafiltration principally reject solids; reverse osmosis and nanofiltration remove ions. Microfiltration or ultrafiltration reject is relatively easy to deal with; it can be recycled back to the head of the process. Microfiltration and ultrafiltration can be an effective pretreatment for nanofiltration or reverse osmosis.

Microfilters and ultrafilters typically are arranged in a hollow-fiber configuration, but flat-sheet configuration is available. Influent is fed into a vessel or tank around the outside of the membrane fibers or inside the fibers. When out-in type membranes are used, product water is collected inside the fibers or at one end of the vessel (for pressure configuration) as permeate. When in-out type membranes are used, product water is collected outside the fibers. Reject water is discharged from the vessel around the outside of the membranes. When large particles can carry over from the secondary process, suppliers recommend fine screens. The membranes must use about 500-μm screens to protect membranes. Practically all vacuum microfiltration and ultrafiltration suppliers scour the outside with air to dislodge solids. Feed water is then used to flush accumulated solids out of the vessel. Effluent turbidity levels consistently less than 0.1 nephelometric turbidity unit (NTU) may be achieved via microfiltration and ultrafiltration.

2.4 Ion Exchange

In ion exchange, ions electrostatically held to charged functional groups on a solid resin surface are exchanged for similarly charged ions in a solution passing across the surface. It is considered a sorption process because the exchange involves transferring ions from a solution phase to a solid phase.

Ion exchange typically is used to soften domestic water by removing "hardness" ions [e.g., calcium (Ca^{2+}) and magnesium (Mg^{2+})]. Wastewater treatment applications include removing nutrients and metals. Ion-exchange deionization typically is used to prepare boiler feedwater and pharmaceutical- and electronics-grade process water.

Ion exchange resins can be made of natural or synthetic materials, although synthetic materials are more common today because of their durability. They are labeled either "strong" or "weak" and either "cation" or "anion," depending on resin strength, polarity, and ability to exchange certain ions and organics.

Cation-exchange resins are negatively charged materials that attract positive ions (cations). Strong acid cation exchangers operate in the pH range of 1 to 14 and will exchange most cations. They exchange hydrogen ions for cations (e.g., calcium) or sodium ions for hardness ions (e.g., calcium and magnesium). Strong acid cation exchangers are regenerated using hydrochloric acid or sodium chloride solutions.

Weak acid cation exchangers operate in the pH range of 7 to 10 and only exchange cations associated with weak anions. They are useful for carbonate hardness removal but not for noncarbonate hardness. Their advantage is that they exchange hydrogen ions (H^+), so they do not add sodium to the water.

Anion exchange resins are positively charged resins that attract negative ions (anions). Strong base anion exchangers operate in the pH range of 1 to 13. They are useful for nitrate removal and exchange hydroxide or chloride for nitrate ions.

Ion exchange resins should be regenerated when the bed is nearly exhausted and breakthrough of specific contaminants is imminent. They are regenerated in co-current or countercurrent modes. In the co-current mode, the regenerant cycle flows in the same direction as the flow during the service cycle (typically downward). In the countercurrent mode, the service cycle typically flows downward and regeneration flows upward.

Countercurrent regeneration is often advantageous in reducing ion leakage, with lower regeneration levels and rinsewater requirements because the ion exchange resin last contacted by the treated water has been regenerated most completely.

2.5   Electrodialysis and Electrodialysis Reversal

Electrodialysis is a membrane process driven by electric current rather than pressure. It typically is used to treat brackish water with a low TDS concentration. Electrodialysis only removes ionized compounds; neutral dissolved organic compounds and suspended solids (e.g., particles and pathogens) pass through them and must be removed by other means.

Unlike reverse osmosis membranes, electrodialysis membranes act depending on the polarity of the electrode where they are placed: anion membranes transfer negative ions; cation membranes transfer positive ions. The ion transfer is driven by a direct-current (DC) electrical field. When direct current is applied to two electrodes submersed in water, an electrical charge is transferred through the liquid via the ionic species in solution (mainly dissolved salts). One electrode becomes the cathode, which is negatively charged. The other becomes the anode, which is positively charged. Because un-like charges attract, the solution's positively charged cations migrate toward the cathode, while the negatively charged anions migrate toward the anode.

Electrodialysis systems typically require chemical addition to prevent membrane surfaces from scaling and fouling. Chemical feed systems may result in high operating and maintenance costs.

An alternative to unidirectional electrodialysis with chemical feed is an electrodialysis reversal (EDR) system. In this system, the electrodes' polarity is reversed three to four times an hour to reduce membrane scaling and fouling by reversing the direction of ion movement in the membrane stack. It typically is designed to operate without constant chemical addition because no flow compartment in the membrane stack is exposed to high solution concentrations for more than 15 to 20 minutes at a time. Buildup of precipitated salts is dissolved and carried away when the cycle reverses.

2.6   Reverse Osmosis and Nanofiltration

Reverse osmosis and nanofiltration use pressure to drive water through the membranes. Natural osmosis occurs when two liquids with different ionic concentrations are separated by a semipermeable membrane. To achieve equilibrium, the membrane allows "pure" water to flow from the low-concentration solution to the high-concentration solution. When the concentrations of both solutions are equal, the flow of water through the membrane will be at equilibrium, and the liquids on either side of the membrane will have different elevations. The difference between the two elevations is the original solution's *osmotic pressure*.

Reverse osmosis occurs when a pressure greater than the osmotic pressure is applied to a solution bound by a semipermeable membrane. The membrane acts as a barrier to solutes. The pressure will drive "pure" water from the more concentrated solution to the other side of the membrane. Permeate (product water) passes through the membrane with reduced solute concentrations, whereas the concentrate (reject) increases solute concentration.

In the current practice of reverse osmosis and nanofiltration, membranes typically are configured as spiral-wound elements (Figure 4.1). Flat membrane sheets are used to construct a spiral-wound unit. A porous membrane support is placed between two membrane sheets, which are oriented so the membrane surface faces out. Three edges of the membrane are glued together to form an envelope, where product water collects. The open end of each envelope is glued to a central collection tube. Holes drilled in the tube allow permeate to flow out of the membrane envelope and into the collection tube. Several membrane envelopes can be attached to a product tube. An open-feed channel spacer is placed over the membrane surfaces. All of the attached membranes then are wrapped around the product tube to form a membrane cartridge. The spiral-wound cartridge fits into a pressure vessel, which can hold several cartridges in series.

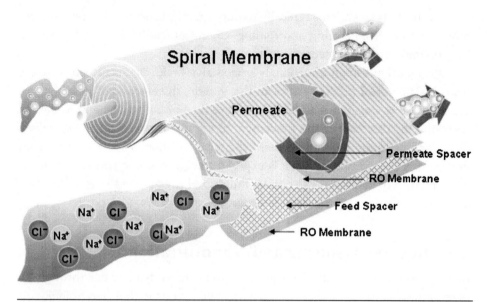

FIGURE 4.1   Spiral-wound membrane module (courtesy of Val S. Frenkel, Ph.D., Kennedy/Jenks Consultants).

Feed water enters one end of the cartridge and flows into the end of the membrane feed-channel spacer parallel to the product water tube. Concentrate (reject) is discharged from the opposite end of the cartridge. The water that passes through the membranes flows through the membrane envelope to the product tube and out either end of the cartridge.

Hollow-fiber reverse osmosis membranes, while less prevalent than spiral-wound elements, also are used in different parts of the world (although typically not for water reuse applications).

Salt (TDS) and TOC rejection vary among membranes. Most of the key manufacturers provide reverse osmosis elements with a single-element salt rejection of more than 99.8% under standard conditions. Feed- and product-water-quality goals govern the design rejection characteristics.

Membrane rejection of different ionic species also varies. Ion rejections typically range from 94% to more than 99%. Large organic molecules are highly rejected. Low-molecular-weight organic compounds often can be removed, but some trace organics [e.g., phenols and certain disinfection byproducts (chloroform, N-nitrosodimethylamine)] typically are not highly removed. The rejection of salt and organics is a function of membrane and solute properties. Solute parameters that primarily affect rejection include molecular weight, molecular size (length and width), acid dissociation constant (pK_a), hydrophobicity/hydrophilicity (log K_{ow}), and diffusion coefficient (D_p). Membrane properties that affect rejection include MWCO, pore size, surface charge (measured as zeta potential), hydrophobicity/hydrophilicity (measured as contact angle), and surface morphology (measured as roughness). Feed-water composition (e.g., pH, ionic strength, hardness, and organics) also influenced solute rejection (Bellona et al., 2004).

Membranes are classified according to the size of solutes/particles removed (Figure 4.2). As shown, reverse osmosis membranes provide the most particle and solute separation.

Nanofiltration membranes have a lower operating pressure than reverse osmosis membranes, so they are less restrictive but provide virus removal and good rejection of many dissolved inorganics, TOC, and trace organics. One drawback, however, is their limited rejection of nitrate. The choice between nanofiltration and reverse osmosis membranes depends on product-water goals and feed-water quality.

Reverse osmosis membranes can be made of several materials. Cellulose acetate was the first compound used, and these membranes had significant drawbacks: hydrolysis-related instability, membrane compaction, and potential for microbial attack. So, various modifications have evolved, including cellulose triacetate and blends of cellulose diacetate–triacetate. The blended diacetate-triacetate membranes were developed to take advantage of cellulose diacetate's superior strength and chemical resistance and cellulose triacetate's high salt rejection. The resulting membranes have better resistance to biological and chemical degradation, improved salt rejection, and longer effective life.

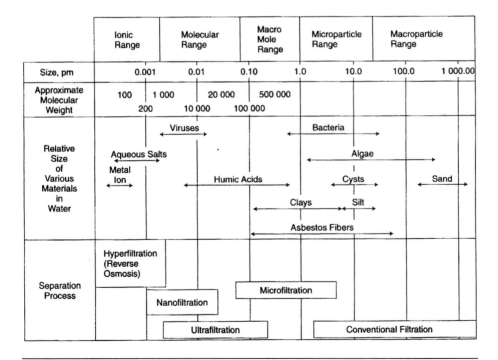

FIGURE 4.2   Membrane classification.

Membranes also are designed as thin-film composites (TFC) with a polyamide support matrix. Polyamide membranes can be operated at lower feed pressure than cellulose acetate membranes but may be damaged by oxidizing via disinfection residuals.

To prevent membrane fouling, proper pretreatment of feed water is essential. A designer may choose the optimum membrane system, but all efforts may be futile if the feed water is not carefully analyzed and appropriately pretreated to protect membranes from sealing or fouling. At a minimum, pretreatment may include a cartridge filter. Feed water with a high volume of suspended solids and organics may require coagulation, sedimentation, conventional filtration, or microfiltration/ultrafiltration. Feed water with significant microbial activity may require disinfection to prevent fouling (polyamide membranes, however, have a low tolerance for chlorine).

Once fouling occurs, membranes must be cleaned. Thorough cleaning of membranes will remove most foulants and prevent irreversible fouling from occurring. Most membrane systems have hard-piped, clean-in-place systems. The cleaning frequency serves as a general guide for evaluating the pretreatment system's effectiveness. A cleaning frequency of more than once a month may indicate inadequate pretreatment.

One major consideration when designing reverse osmosis or EDR plants is disposing of the concentrate. The approach depends on the plant location (e.g., coastal or inland state).

2.7   Chemical Oxidation

Chemical oxidation processes convert undesirable chemical parameters to ones that are neither harmful nor objectionable. A chemical substance is oxidized when it loses electrons to another substance (the oxidizing agent). The loss of electrons increases the substance's oxidation state (valence). Chemical oxidation often is used to control inorganic (e.g., Mn^{2+}, Fe^{2+}, S^{2-}, CN^-, SO_3^{2-}) and organic (e.g., phenols, amines, humic acids, toxic compounds, organic contaminants of emerging concern, bacteria, and algae) compounds.

Thermodynamic principles can be used to determine whether specific oxidation reactions are possible. In most cases, it is impractical to completely oxidize a chemical. Depending on the oxidant and oxidizing conditions, the intermediate oxidation products are much less toxic compounds with less-objectionable characteristics (Weber, 1972), so further oxidation often is not required and, from an economic standpoint, may not be justified. Designers also should realize that reaction times for complete oxidation are often so long that designing facilities with sufficient detention times is impractical. It is a common practice in reclaimed water treatment to define *chemical oxidation* as a method for selective inactivation or modification of toxic or objectionable substances.

The intended use of reclaimed water may restrict the type of oxidizing agents used because of residual toxic or other deleterious effects. Ideally, no oxidant residue should remain after treatment is completed, particularly when augmenting potable water supplies. Other significant aspects that should be considered when selecting suitable oxidizing agents for reclaimed water treatment include treatment effectiveness, cost, ease of handling, compatibility with preceding or subsequent treatment steps, and nature of the oxidation operation.

Only a few oxidizing agents can meet most of these requirements: oxygen (air), ozone, hydrogen peroxide, potassium permanganate, chlorine (or hypochlorites), and chlorine dioxide. Combinations of chemicals (e.g., hydrogen peroxide and ozone) are sometimes more effective oxidants; such processes are called *advanced oxidation processes*. In general, given sufficient effectiveness, the choice of oxidizing agent is based largely on economics and the material's handling characteristics.

The efficiency of oxidation reactions depends on the target compound. For example, using ozone to oxidize inorganic constituents of wastewaters is a rather rare application, because other methods exist for most of these target compounds (ozone may be used for residual color removal). However, inorganic compounds [e.g., manganese-(II) and iron-(II)] may be oxidized as a secondary effect of ozonation for other purposes.

There is an increasing interest in implementing oxidation processes to remove organic microconstituents, most of which are only poorly accessible to a direct ozone attack. In almost all cases, the target compounds will not be mineralized but transformed to metabolites, which typically are more polar and smaller in molecular weight. Quite often, some of the products formed do not react further with ozone. Because complete removal of organic products does not occur, it is essential to have a subsequent treatment process in place (e.g., biological filtration systems or an activated carbon contactor).

2.8   Advanced Oxidation Processes

Advanced oxidation processes (AOPs) are oxidation processes that generate highly reactive intermediates (e.g., radicals, especially OH radicals). Ozone alone at high pH can be considered an AOP; it reacts with most organic contaminants in natural waters primarily via a nonselective indirect pathway. Advanced oxidation processes are alternative techniques for catalyzing the production of radicals, thereby accelerating the destruction of organic contaminants. Because the radicals are relatively nonselective in their mode of attack, they can oxidize all reduced material, not just specific classes of contaminants (as is the case with molecular ozone).

Because of ozonation's slow reaction-rate constants and mostly incomplete mineralization, treatment methods with an even stronger oxidant (the OH^- radical) were developed:

- Ozonation at high pH,
- Ozone and hydrogen peroxide (O_3/H_2O_2),
- Ozone and UV radiation (O_3/UV),
- UV radiation and hydrogen peroxide (UV/H_2O_2),
- Hydrogen peroxide and chlorine, and
- Ozonation in the presence of a solid catalyst (e.g., titanium dioxide).

Combining two oxidants increases the oxidation potential; the treatment can be more successful than that of a single oxidant (UV, H_2O_2, or O_3).

Advanced oxidation is used in reclaimed water treatment to remove microconstituents and nitrosamines, especially N-nitrosodimethylamine (NDMA). (For further reference, staff at the Orange County, California, Ground Water Replenishment project conducted a number of studies on the effectiveness of AOPs in removing 1,4-dioxane and NDMA.) To further remove organics, AOPs can be followed by biological filtration.

2.9   Activated Carbon Adsorption

In granular activated carbon treatment, water typically is passed through fixed or semifixed beds that contain granular activated carbon. The carbon has an extremely high specific surface area that preferentially adsorbs nonpolar organic molecules, removing them from the water. Inside each granular activated carbon filter, concentration profiles develop for one adsorbing component, and after the minimum contact time elapses, an active transport zone develops where the adsorbing molecules migrate from the water onto the carbon. This transport zone slowly migrates through the bed until the front of the zone reaches the end of the filter, and breakthrough begins. The granular activated carbon behind the active transport zone can no longer remove the contaminant from water (i.e., is saturated). Once the transport zone leaves the filter, the effluent concentration will approach the influent concentration.

In single-filter operations, contactor inflow typically is diverted before the effluent concentration no longer meets the minimum treatment objective, and the granular activated carbon is regenerated. Proper design of a granular activated carbon process typically involves series, parallel, or countercurrent filter operations to maximize the carbon's adsorption capacity. It maximizes carbon-use rates while maintaining treatment objectives. When the carbon's adsorption capacity has been exhausted, the spent carbon can be either disposed of at landfills or thermally regenerated onsite or offsite in a furnace facility.

Granular activated carbon adsorption primarily is used to remove refractory and residual organics. It also provides some trace inorganic reduction via adsorption of inorganics chelated with organic compounds. In addition, the adsorption beds provide a medium for some fixed-film biological growth that may reduce residual biodegradable dissolved organic carbon and nitrate.

Recently, considerable attention has been focused on enhancing this biological treatment with pre-ozonation. Pre-oxidation processes can oxidize refractory organics to biologically assimilable forms, which can extend the carbon's useful performance between reactivations. This may not help if the granular activated carbon filter is being used as a biologically activated carbon filter, because such filters do not require regeneration. Oxidation processes (e.g., ozone) also provide dissolved oxygen for bacterial respiration. However, pre-ozonation may create inorganic disinfection byproducts (e.g., bromates) that can be problematic in the water supply.

Pretreatment typically includes chemical clarification and filtration of secondary effluent to reduce the solids that could plug or clog granular activated carbon filters.

2.10 Air Stripping

Air stripping, which volatizes and exhausts contaminants in water, is used to remove volatile organic compounds, ammonia, and carbon dioxide.

In air-stripping processes, water is pumped to the top of a tower packed with media, evenly distributed across the media, and allowed to fall downward in a film layer along the packing surfaces. Meanwhile, air blown into the base of the tower flows upward, and as it contacts the large surface area, volatile contaminants transfer from water to air. The volatile contaminants are carried out of the column and into the atmosphere.

The early Water Factory 21 and the Denver Potable Water Reuse Demonstration Plant initially used air stripping. Water Factory 21 decommissioned air strippers once reverse osmosis was introduced. The downside to air stripping is that it transfers water contaminants into the air column, so it may require more time and money for air permitting and treatment issues.

The air-stripping process for ammonia removal consists of raising the water's pH to between 10.8 and 11.5 and circulating large quantities of air through the tower. The conversion of ammonium ion in water to ammonia gas in air is a consequence of the pH dependency of the ammonia–ammonium equilibrium:

$$NH_3 + H_2O \leftarrow NH_4^+ + OH^- \qquad (4.2)$$

Raising the pH significantly higher than the equilibrium point (pH 9.3) shifts the equilibrium to the left, so ammonia is the predominant form. Ammonia is volatile and responds well to air stripping. Because lime or caustic soda is added to raise the pH, air-stripping ammonia works well when preceded by a high-pH lime-clarification process. However, calcium carbonate scaling can occur in the tower and on the structural members, resulting in high maintenance costs. So, biological nitrogen removal (nitrification and denitrification) is often the preferred ammonia-removal process.

Air stripping to remove volatile organic compounds from water relies on the tendency for moderately soluble organic compounds to leave the liquid phase and enter the vapor phase (air). The ratio of gas to liquid is in the range of 100 to 1 (by volume), approximately 10-fold less than that required for ammonia stripping.

Air stripping typically is used after membrane treatment to remove excess carbon dioxide from water. Carbon dioxide is highly volatile and can be readily stripped; removing it raises the water's pH and reduces its corrosive properties.

3.0   NATURAL TREATMENT SYSTEMS

When augmenting potable water sources, natural treatment systems alone may have some limitations in meeting desired water-quality goals. When combined with traditional and other advanced treatment processes, however, natural treatment systems may serve both public perception and treatment goals. Perception goals are met by providing a "natural" separation of time and space between waste treatment and water supply. Treatment goals are met by removing certain contaminants that traditional treatment processes did not treat. The natural treatment systems discussed in this chapter are constructed wetlands, soil–aquifer treatment, and riverbank filtration.

These systems remove certain contaminants via "natural" environmental components [e.g., vegetation, soil, and microorganisms (soil and aquatic)]. So, when selecting a natural treatment system, project designers should consider which contaminants need to be removed or neutralized to meet water-quality goals. Natural systems are limited in their ability to remove constituents and, at certain times of the year, may export or recycle back to the water column (wildlife excretions, phosphorus and nitrogen when plants die off, etc.). Kinetic equations have been developed to determine the background concentrations of biochemical oxygen demand, total phosphorus, total nitrogen, etc. already attributable to the wetland processes.

It is possible that the natural treatment system may deteriorate or alter reclaimed water quality so as not to meet water-quality objectives. In this case, although public perception goals may be met, further treatment may be required.

3.1   Constructed Wetlands

Constructed wetlands are designed to mimic natural wetlands and treat reclaimed water using emergent plants (e.g., cattails, reeds, and rushes). The basic treatment mechanisms include sedimentation, chemical precipitation and adsorption, microbial interactions with organic matter and nitrogen, and some uptake by vegetation. Wetland systems are typically anoxic and preclude significant nitrification, but they allow denitrification if upstream nitrification occurs. They can reduce concentrations of organic carbon, suspended solids, nitrogen, significant levels of metals, certain trace organics, and pathogens. Phosphorus-removal capability varies depending on site-specific factors, especially soil type.

There are two types of constructed wetlands: free water surface systems and subsurface flow systems. In free water surface systems, water flows at shallow depths [0.3 to 0.9 m (1 to 3 ft)] in constructed basins or channels over a soil surface. The soil surface can be lined to prevent seepage to groundwater. In a subsurface soil system wetland, water flows below the surface in a trench or bed filled with porous media (typically gravel). Both systems use emergent aquatic vegetation to provide microbiological reactions for treatment.

The choice of system depends on such factors as economics, land availability, and treatment objectives. The media in subsurface flow system wetlands typically provide a larger surface area for treatment, so subsurface flow system wetlands need less land than free water surface wetlands to treat a given amount of flow. Subsurface flow system wetlands also have more thermal protection in cold climates and fewer odor, mosquito, and other nuisance problems because the water is underground. Such advantages, however, are offset by the high cost to procure, deliver, and place gravel media (Reed et al., 1988) and the limited opportunity for benefits other than water-quality improvement. Free water surface wetlands, on the other hand, frequently are designed to maximize wetland habitat values and reuse opportunities while improving water quality (U.S. EPA, 1993).

3.2 Soil–Aquifer Treatment

Recharging groundwater with reclaimed water is an increasingly valued practice for augmenting potable water supplies, especially in arid areas. Reclaimed water groundwater recharging can be introduced through the use of recharge basins, injection wells, and alluvial filtration.

The most widely used method of groundwater recharge is spreading water into recharge basins. Recharge basins are shallow earthen ponds underlain with permeable soil that allows reclaimed water to infiltrate the soil surface and percolate through the vadose zone into the underlying aquifer. *Soil–aquifer treatment* is the supplemental treatment of reclaimed water that occurs during percolation through the soil and groundwater transport and storage (Figure 4.3).

Many SAT systems have been or are scheduled to be installed in the United States and other countries because they are low-maintenance and economical where land is available. Los Angeles County implemented SAT systems in

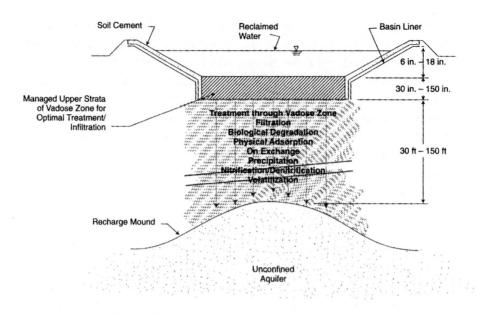

FIGURE 4.3   Soil–aquifer treatment percolation basin (in. × 25.4 = mm; ft × 0.304 8 = m).

Montebello Forebay in the early 1960s, and research there showed that the risks associated with using reclaimed water for groundwater recharge were not significantly different than those associated with using natural groundwater with no reclaimed water additions. Reclaimed water had no measurable effect on groundwater quality or human health (Nellor et al., 1984). Another epidemiologic study (Sloss et al., 1996) concluded that, after almost 30 years of recharging groundwater with reclaimed water, rates of cancer, mortality, and disease in the Los Angeles County area that receive reclaimed water are similar to those in areas that do not receive reclaimed water.

A soil aquifer treatment system depends on reclaimed water infiltrating the soil and moving away from the recharge basin. Water quality can be improved via the many mechanisms that occur in soil (e.g., filtration, biological degradation, physical sorption, ion exchange, and precipitation). These mechanisms are effective in removing organic carbon, nitrogen, phosphorus, suspended solids, pathogens, trace metals, and trace organic compounds. Removal efficiencies for specific water-quality constituents depend on the type of soil, level of pretreatment, loading rate, loading cycle, and temperature (Table 4.2). The removal of bulk and trace organics and nitrogen can be a continuous, sustainable process that relies primarily on biodegradation. The removal of trace metals and phosphorus via precipitation or ion exchange may result in accumulation of phosphorus or metals in the soil, but estimates show that it would take decades for accumulated trace metals to become a problem. Moreover, the accumulation can

TABLE 4.2 Contaminants removed via advanced treatment processes.

	Suspended solids	Pathogens	Phosphorus	Nitrogen	Metals	Total organic carbon	Volatile organic compounds	Trace organics	Total dissolved solids
Biological nutrient removal	N/A	None	Good	Good	Moderate	Good	Moderate	Moderate	None
Coagulation/flocculation (with sedimentation)	Good	Poor	Moderate	N/A	Good	Moderate	N/A	N/A	None
Microfiltration and ultrafiltration	Good	Good[a]	Poor[b]	Poor[b]	Poor[b]	Poor[b]	Poor[c]	Poor[b]	None
Membrane bioreactor	Good	Good	Moderate	Moderate	Moderate	Good	Moderate	Moderate	None
Reverse osmosis	N/A	Good	Good	Good	Good	Good	None[d]	Good	Good
Nanofiltration	N/A	Good	Good	Good	Good	Good	None[d]	Moderate	Moderate
Ion exchange	Poor	None	Good	Good	Moderate	Poor	None	Moderate	None[e]
Electrodialysis/electrodialysis reversal	None	None	Good[f]	Good[f]	Moderate	Moderate	Poor	None	Good
Chemical oxidation (ozone)	None	Good[g]	Poor	Poor	Poor[h]	Moderate	Good	Good	None
Advanced oxidation processes	None	Good	None	Poor	Poor	Moderate	Good	Good	None
Medium-pressure UV irradiation	None	Good	None	None	None	Poor	Poor	Moderate	N/A
Activated carbon	Moderate	Poor	Poor	N/A	N/A	Good	Poor	Moderate	N/A
Air stripping	N/A	N/A	N/A	Poor[i]	N/A	Poor	Good	Poor	N/A
Natural treatment systems	Good	Moderate	Moderate	Moderate	Moderate	Moderate	Moderate	Moderate	N/A

[a] Ultrafiltration has smaller pores than microfiltration and has better pathogen removal. Ultrafiltration highly removes virus and microfiltration does not.
[b] Microfiltration/ultrafiltration typically is poor for nutrient, metal, total organic carbon, and trace organic removal but can be increased to moderate with coagulant addition in direct filtration mode.
[c] For submerged vacuum-type system and none for pressure vessel-type systems.
[d] Volatile organic compounds can be removed in post-treatment processes (i.e., air strippers).
[e] For cation-anion demineralization systems, TDS removal is excellent (but not economical for high-TDS waters).
[f] Removes charged ions. No removal of nutrients in organic form (e.g., total Kjeldahl nitrogen).
[g] Ozone effectiveness on pathogens is limited in cold temperatures.
[h] Other processes are needed to obtain removal.
[i] Excellent ammonia removal in ammonia stripping towers.
TDS = total dissolved solids.

be removed by taking off a relatively thin layer of soil, thereby restoring the ground to near-original condition.

Research has demonstrated that removal of bulk organics, nitrogen, and pathogens occurs in the upper strata of soil in the vadose zone (Bouwer et al., 1984; Fox et al., 2001, 2006; Rauch-Williams and Drewes, 2006). Longer retention times typically are required to remove trace organics (Montgomery-Brown et al., 2003; Drewes et al., 2003; Mansell and Drewes, 2004). Research also suggests that a correlation may exist between treatment efficiency and certain predominant soil characteristics in the vadose zone (Bouwer et al., 1984, 2006).

During the course of SAT operations, solids filtration and biofilm development at the basin–soil interface forms a "clogging layer" that can reduce infiltration rates (Bouwer et al., 1980, 1984; Fox et al., 2001, 2006). This reduction in infiltration rates often controls SAT operations, making cyclical flooding and drying of basins necessary. The basins are flooded for a given period, until the clogging layer reduces infiltration rates. Then basin flooding ceases, the water remaining in the basin is allowed to infiltrate, and then the basin's soil surface is allowed to dry. The clogging layer is removed either by natural processes (e.g., desiccation and wind) or by physically cleaning the soil surface.

Cycle times also influence the depth of oxygen levels in the vadose zone, so they are critical in optimizing biological removal (in particular, nitrogen removal when un-nitrified effluents are applied) and preventing reduced infiltration rates.

When siting SAT systems, it is important to conduct a detailed hydrological and geological analysis of the proposed site. Key factors that make a site suitable are surface infiltration, vadose zone transmission, and aquifer transmissivity and storage. If the selected sites have enough aquifer transmissivity and depth to groundwater, then groundwater mounding should not reduce the amount of reclaimed water that can be recharged. A detailed soil evaluation of soil properties such as texture, structure, and limiting layers should be fully examined to reduce failures and ensure successful design. The loading rates typically used for onsite systems can be used as first approximations for sizing purposes.

3.3   Riverbank Filtration

Riverbank filtration (RBF) is a natural process that has been used for public and industrial water supplies in Europe for more than a century. In the United States, it has been used for nearly half a century to treat a variety of waterborne contaminants and is increasingly recognized as a viable option for water utilities that must meet stringent regulations for the direct use of surface water and "groundwater under the direct influence of the surface" (GWI), both of which may be impaired by wastewater discharges. The U.S. Environmental Protection Agency's Long-term 2 Enhanced Surface Water Treatment Rule (LT2) now recognizes RBF as an effective pretreatment for surface waters and GWIs; it provides removal credit for *Giardia* and *Cryptosporidium* based on the travel distance from the surface water to the extraction well if the aquifer is shown to contain a minimum level of "fine sediments" (U.S. EPA, 2006).

In riverbank filtration, a withdrawal well is installed along the bank of a river or lake, or near a recharge basin. Pumping water from the well establishes a groundwater gradient from the surface water to the well. Adequate travel times or aquifer retention (for at least a few days) is necessary to improve water quality.

While riverbank filtration is well established as an adequate barrier to protozoan cysts and bacteria, little information is available about specific virus-removal efficiencies. One difficulty with RBF field studies is that, to establish virus-removal levels, the raw water must contain a large number of viruses. Because "naturally occurring" levels of viruses are insufficient, most of the available information about RBF's virus removal is based on removing surrogate microorganisms (typically bacteriophages). F-specific ribonucleic acid (FRNA) bacteriophages have been shown to represent a "worst-case" virus model for assessing subsurface transport because of FRNA's poor adsorption to soil particles and high survivability in groundwater when compared to enteric viruses (Havelaar, 1993). And even though there is evidence that RBF reduces viruses, it does not replace the need for disinfection (Ray et al., 2002).

There is also evidence that RBF reduces pathogens, nutrients, total organic carbon, disinfection byproduct precursors, distribution system and storage regrowth potential, and fouling potential for membrane filters. More recently, it has been reported that RBF can effectively attenuate organic microconstituents that might be present in impaired source water, although the understanding of how RBF removes organic and inorganic microconstituents is still being researched.

When locating RBF systems, it is important to conduct a detailed hydrological and geological analysis of the proposed site. Infiltration rates typically do not decline over time (as observed in SAT systems) because the river sediments are the result of high flow velocities or flooding events. So, maintenance is limited to servicing the groundwater abstraction pumps, resulting in low O&M costs.

A significant difference between RBF and SAT is the *hyporheic zone*—the interface between surface water and groundwater (Tufenfkji et al., 2002). It is a transition zone with a distinct biogeochemical environment (characterized by gradients in light, temperature, pH, redox potential, oxygen, and organic carbon) that controls the quality of bank filtrate.

Riverbank filtration is a more active treatment process than SAT is. A riverbank filtration "design" must account for aquifer characteristics that provide established treatment reductions by selecting a site that ensures minimum aquifer retention time. Retention time also is affected by and controlled via the withdrawal well's pumping rate. The design approach that ensures adequate RBF treatment must be based on a well-designed aquifer and groundwater sampling program that relates water quality to hydrogeologic characteristics to well withdrawal rates.

Basically, riverbank filtration is a proven pretreatment for potable surface water and GWI supplies that can provide another protective barrier for reclaimed water used to augment those supplies.

4.0   DISINFECTION

Disinfection processes destroy and inactivate pathogens in water; however, this is not sterilization (the inactivation of all organisms). Disinfection targets bacteria and viruses; parasites (e.g., *Giardia lamblia* and *Cryptosporidium parvum*) are resistant to conventional disinfection methods and must be removed via other treatment processes (e.g., ozonation or microfiltration).

Wastewater disinfection typically is accomplished via chemical means (e.g., chlorination or ozonation), photochemical means (e.g., UV irradiation), or physical means (e.g., membrane filtration). Disinfection effectiveness is measured by an indicator organism because it is impractical to measure each bacterium, virus, and parasite in reclaimed water. There is no universally accepted indicator organism for measuring reclaimed water's microbiological quality. *E. coli* has replaced fecal coliform to a large extent in Canada; total coliforms are measured less and less because of their ubiquitous presence.

The California reuse regulations have four major categories for the reuse water, based on application and type of treatment. Three categories require disinfection and limit total coliforms to up to 2.2 most probable number (MPN)/ 100 mL (7-day median), depending on the potential for human contact and ingestion. Florida requires disinfection and no detectable fecal conforms for 75% of samples over a 30-day period (U.S. EPA, 2004). Arizona requires nondetect in 4 out of 7 samples for fecal coliform with a not to exceed (NTE) level of 23 CFU/100 mL (Arizona Administrative Code Title 18, Chapter 9).

4.1   Chlorination

Chlorination is the most common method of disinfection in the United States because of its low cost, reliability, and detectable residual. One of its disadvantages, however, is the potential formation of carcinogenic compounds (e.g., trihalomethanes) caused by chlorine reacting with organics in wastewater. Also, depending on where the reclaimed water will be discharged, dechlorination may be necessary. Wetlands, for example, would be negatively affected by relatively high concentrations of chlorine.

Chlorine disinfection's effectiveness is a function of the product of contact time and chlorine residual. There is no universal agreement on the contact time required to completely remove pathogens. The Pomona Virus Study (County

Sanitation Districts of Los Angeles County, 1977) demonstrated an approximate 5-log removal for viruses using 2-hour contact times and chlorine residuals of 5 to 10 mg/L. Operating experience at Los Angeles County Sanitation District plants has since indicated that virus removal can be achieved with lower residuals and contact times, although 5-log removal has not been documented (Yanko, 1993). Florida and Arizona reuse standards require 1 mg/L chlorine residual after a 15-minute contact time at peak hourly flows (U.S. EPA, 2004).

Chlorine is extremely hazardous in gaseous form, so adequate safety and risk-management program measures must be included in the gaseous chlorination system design and operations to prevent or control toxic gas leaks. Because of these safety concerns, many communities use liquid chlorine compounds (e.g., sodium hypochlorite, which has relatively high costs). Onsite generation of chlorine is another viable option.

4.2   Ozone Disinfection

Ozone disinfection is becoming an increasingly popular alternative to chlorination. Ozone is a powerful oxidant and therefore an effective disinfectant, especially for *Giardia* and *Cryptosporidium*. It has similar bactericidal properties to chlorine and is equal or superior to chlorine in its ability to inactivate viruses (Metcalf and Eddy, 2005).

Ozone is produced when a high voltage is imposed on a discharge gap in the presence of a gas containing oxygen. Because ozone is a relatively unstable gas, it must be generated onsite from air or pure oxygen. The pure oxygen may be obtained from an oxygen-generation system that allows for oxygen to be recycled from the ozone contact chamber back to the generation system. When ozone is added to water, it rapidly reverts to oxygen, so no chemical residual persists in the treated water. Ozone decomposes in water by forming hydrogen peroxy (HO_2) and hydroxyl (OH) free radicals. These free radicals have a great capacity for destroying bacteria via cell wall disintegration (cell lysis) and virus inactivation by damaging nucleic acid constituents and breaking carbon nitrogen bonds. If a disinfectant residual is required, chlorine may be added as a postdisinfectant.

Although ozone disinfection primarily has been used in the water treatment field, recent advances in ozone generation and solution technology have made it more cost-effective for reclaimed water. Ozone also can be used to control odors and remove soluble refractory organics.

Recently, a number of compounds have been identified as ozonation byproducts (e.g., bromate and aldehyde), and some could have adverse health effects. For example, bromate (which is formed when ozone oxidizes bromide in water) is now considered a genotoxic carcinogen. The revised World Health Organization guidelines for drinking water indicate that the bromate limit should be

25 µg/L. This value was set to account for some of the analytical problems associated with measuring bromate (WHO, 2004). The U.S. Environmental Protection Agency has proposed a bromate maximum contaminant level (MCL) of 10 µg/L (U.S. EPA, 1994).

4.3   Ultraviolet Irradiation

Ultraviolet irradiation has increasingly been used to disinfect treated reclaimed water in recent years, so the design and effectiveness of UV disinfection systems have been scrutinized. If designed and maintained properly, UV systems can inactivate coliforms and viruses effectively. The National Water Research Institute has published a set of UV disinfection guidelines (used as the design guide in California) that should be referenced as needed.

Ultraviolet irradiation, which produces energy in a particular wavelength range, damages the genetic material of organisms. The source of this radiation is a low-pressure, medium-pressure, or high-pressure lamp emitting light at 253.7-nm wavelengths, which is within the optimum germicidal wavelength range (250 to 270 nm).

Studies have shown that photoreactivation and dark repair can occur after UV inactivation (Oguma et al., 2001; Zimmer and Slawson, 2002), so using UV as the only disinfection method can lead to a gradual recovery of certain organisms. (A detailed discussion of photoreactivation and dark repair is beyond the scope of this publication but should be researched further by utilities interested in UV disinfection.)

Most ultraviolet irradiation system designs consist of a reactor (closed vessel or open channel) with a number of lamps submerged in the liquid to be treated. The lamps may be parallel or perpendicular to the flow. Each lamp is encased in a quartz tube. As the liquid flows through the lamp array, it is irradiated. In another design, flow is conveyed in polytetrafluoroethylene tubes interspersed among the lamps so the flow is parallel to the lamps not in contact with them. (Polytetrafluoroethylene is transparent to UV irradiation.)

The radiation dose is determined by radiation intensity and residence time in the reactor. An ultraviolet irradiation system's effectiveness at supplying the necessary disinfection dose depends on many factors [e.g., hydraulic design, absorbance by the liquid, presence of particulates, transmittance of the lamps and quartz tubes, and the presence of iron and nitrates (high levels of nitrate, in particular, because UV energy is lost)]. Upstream iron addition for any pretreatment can adversely affect UV scaling.

Because radiation intensity diminishes with distance, the liquid and lamps should be in intimate contact at all points in the reactor. To maximize the opportunity for all organisms to receive the required radiation dose, there should be equal contact between the lamps and all units of liquid. This is best achieved by both maintaining plug flow through the reactor to maximize contact time and creating enough turbulence to cause transverse dispersion, which maximizes radiation exposure. Hydraulic short-circuiting should be avoided because it shortens residence time and loses energy, resulting in less-effective disinfection.

The quality of reclaimed water to be disinfected has a large effect on radiation intensity. When designing a UV system, engineers should take into account the reclaimed water's UV demand (quantified by its absorbance at a wavelength of 253.7 nm), the concentration of suspended solids, and particle size distribution. Lamp transmittance, which also affects intensity, is a function of lamp age and the external and internal condition of the quartz tube. Buildup of bacteria, algae, scale, or other deposits reduces lamp transmittance, so the amount of lamp maintenance and likelihood of lamp fouling should be anticipated. Also, lamp intensity should be monitored so lamps requiring cleaning or replacement may be identified before disinfection-dose losses occur.

5.0   TREATMENT TRAINS

The ability of the previous processes to remove contaminants, especially TDS, TOC, and microbiological matter, is a key consideration when developing and evaluating a treatment train. Numerous combinations of treatment trains can be developed to meet the requirements for reclaiming water so it can be used to augment potable water sources. Figure 4.4 provides examples.

6.0   CAPITAL AND OPERATING COSTS

Capital and operating costs are critical components of all projects that affect the public. The costs of an indirect potable reuse project will depend on the treatment processes selected.

Project costs should be estimated at the following stages:

- When the project is conceived (before any money has been spent);
- During planning, while feasibility studies or facility plans are being prepared and the best two or three alternatives must be selected;
- During design, when the most cost-effective plan must be chosen; and
- After plans are completed, as a basis for informing the owner of probable cost and judging bids.

One critical cost consideration is the probable cost of the selected project versus that of developing another source of supply (not reclaimed water). So, project costs should be reviewed and modified throughout the project's life.

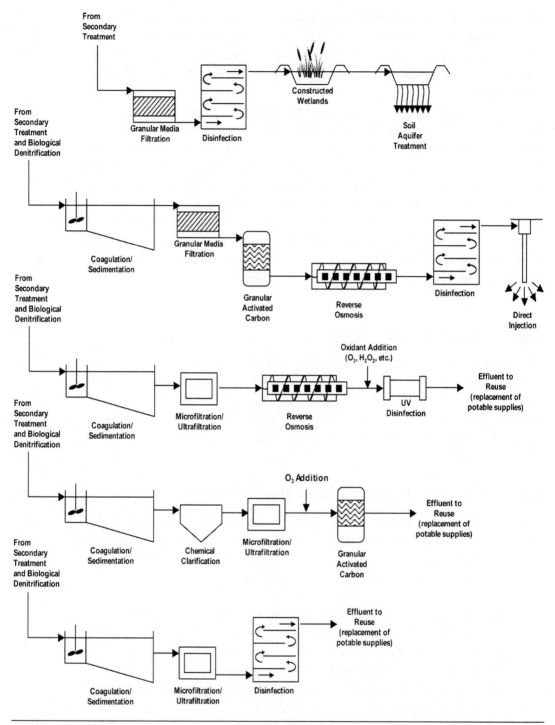

FIGURE 4.4   Examples of treatment process trains for indirect reuse.

7.0 SUMMARY

Many treatment processes may be used to meet an indirect potable reuse project's water-quality objectives. When selecting a treatment combination, it is important to follow the concept of multiple barriers. Pilot-scale testing and research also may be performed before design to ensure that water-quality objectives can be met at a reasonable cost.

Pilot testing typically is done for the following reasons:

- Competitive comparison—this approach is used when several membranes or processes qualify for a given set of treatment objectives, or when a new product is introduced to the market.
- Owner request—even though a selected process is well known, the owner may request a demonstration to meet public perception goals or to help operations staff gain confidence.
- Regulatory needs—some regulators may be unfamiliar with a process and, therefore, require pilot testing before permitting the project.
- Proof-piloting—this approach is used to verify manufacturer claims.

Pilot testing may not be required if the selected process is well understood, consistently meets reuse water-quality goals, and there is adequate information on long-term operating performance (both costs and water quality). In some areas, historical data can be used to make decisions on reuse water-quality results and lifecycle cost effects. These upfront decisions typically are project- and site-specific.

8.0 REFERENCES

Bellona, C.; Drewes, J. E.; Xu, P.; Amy, G. (2004) Factors Affecting the Rejection of Organic Solutes during NF/RO Treatment—A Literature Review. *Water Res.*, **38**, 2795–2809.

Bouwer, E. J.; McCarty, P. L.; Bouwer, H.; Rice, R. C. (1984) Organic Contaminant Behavior during Rapid Infiltration of Secondary Wastewater at the Phoenix 23rd Avenue Project. *Water Res.* (G.B.), **18**, 463.

Bouwer, H.; Rice, R. C.; Lance, J. C.; Gilbert, R. G. (1980) Rapid-Infiltration Research—The Flushing Meadows Project, Arizona. *J. Water Pollut. Control Fed.*, **52** (2), 457.

Bratby, J. R. (1980) *Coagulation and Flocculation*; Uplands Press Ltd.: London.

Bratby, J. R. (1996) An Update on Nutrient Removal in Wastewater Treatment. *Proceedings of the International Seminar on Wastewater Treatment*; Brasilia, Brazil.

County Sanitation Districts of Los Angeles County (1977) *Pomona Virus Study*; Whittier, California.

Drewes, J. E., Heberer, T.; Rauch, T.; Reddersen, K. (2003) Fate of Pharmaceuticals during Groundwater Recharge. *J. Ground Water Monitoring Remediation*, **23** (3), 64–72.

Fox, P.; Houston, S.; Westerhoff, P.; Nellor, M.; Yanko, W.; Baird, R.; Rincon, M.; Gully, J.; Carr, S.; Arnold, R.; Lansey, K.; Bassett, R.; Gerba, C.; Amy, G.; Reinhard, M.; Drewes, J. E. (2006) *Advances in Soil–Aquifer Treatment for Sustainable Water Reuse*; American Water Works Association Research Foundation: Denver, Colorado.

Fox, P.; Nellor, M.; Drewes, J. E.; Houston, S.; Westerhoff, P.; Muccino, J.; Arnold, R.; Gerba, C.; Wilson, L. G.; Lansey, K.; Bassett, R.; Amy, G.; Reinhard, M.; Baird, R.; Yanko, W.; Rincon, M. (2001) *An Investigation of Soil–Aquifer Treatment for Sustainable Water Reuse*; American Water Works Association Research Foundation: Denver, Colorado.

Havelaar, A. H. (1993) Bacteriophages as Models of Human Enteric Viruses in the Environment. *ASMNews*, **59**, 614–619.

Mansell, J.; Drewes, J. E. (2004) Fate of Steroidal Hormones during Soil–Aquifer Treatment (SAT). *J. Ground Water Monitoring Remediation*, **24** (2), 94–101.

McEwen, B.; Richardson, T. (1996) Indirect Potable Reuse: Committee Report. *Proceedings of the American Water Works Association and Water Environment Federation Water Reuse Conference*; San Diego, California, Feb; pp. 486–503.

Metcalf and Eddy, Inc. (2005) *Wastewater Engineering: Treatment, Disposal, Reuse*; 4th ed.; McGraw-Hill: New York.

Montgomery-Brown, J.; Reinhard, M.; Drewes, J. E.; Fox, P. (2003) Behavior of Alkylphenol Polyethoxylate Metabolites during Soil–Aquifer Treatment. *Water Res.*, **37**, 3672–3681.

Nellor, M.H., Baird, R. B.; and Smyth, J. R. (1984) *Health Effects Study Final Report*; County Sanitation Districts of Los Angeles County: Whittier, California.

Oguma, K., Katayama, H.; Mitani, H.; Shigemitsu, M.; Hirata, T.; and Ohgaki, S. (2001) Determination of Pyrimidine Dimers in *Escherichia coli* and *Cryptosporidium parvum* during UV Light Inactivation, Photoreactivation, and Dark Repair. *Appl. Environ. Microbiol.*, **67** (10), 4630–4637.

Rauch-Williams, T.; Drewes, J. E. (2006) Using Soil Biomass as an Indicator for the Biological Removal of Effluent-Derived Organic Carbon during Soil Infiltration. *Water Res.*, **40**, 961–968.

Ray, C.; Melin, G.; Linsky, R. B. (2002) *Riverbank Filtration—Improving Source-Water Quality*; National Water Research Institute: Fountain Valley, California; Kluwer Academic Publishers: Norwell, Massachusetts.

Reed, S.C.; Middlebrooks, E. J.; Crites, R. W. (1988) *Natural Systems for Waste Management and Treatment*; McGraw-Hill: New York.

Sloss, E. M.; Geschwind, S.; McCaffrey, D. F.; Ritz, B. R. (1996) *Groundwater Recharge with Reclaimed Water, An Epidemiologic Assessment in Los Angeles County, 1987–1991*; RAND: Santa Monica, California.

Tufenfkji, N.; Ryan, J. N.; Elimelech, M. (2002) The Promise of Bank Filtration. *Environ. Sci. Technol.*, **36**, 422a–428a.

U.S. Environmental Protection Agency (1993) *Constructed Wetlands for Wastewater Treatment and Wildlife Habitat*; EPA-832/R-93-005; Washington, D.C.

U.S. Environmental Protection Agency (1994) *Draft D/DBP Rule Language*; Washington, D.C.

U.S. Environmental Protection Agency (2004) *Guidelines for Water Reuse*; EPA-625/R-04-108; Technology Transfer: Washington, D.C.

U.S. Environmental Protection Agency (2006) *Draft LT2EsWTR Implementation Guidance*; Washington, D.C.

Weber, W. J. (1972) *Physiochemical Processes for Water-Quality Control*; Wiley & Sons: New York.

World Health Organization (WHO) (2004) *Guidelines for Drinking Water Quality*, 3rd Ed., Vol. 1.,

Yanko, W. A. (1993) Analysis of Ten Years' Virus Monitoring Data from Los Angeles County Treatment Plants Meeting California Wastewater Reclamation Criteria. *Water Environ. Res.*, **65** (3), 221.

Zimmer, J. L.; Slawson, R. M. (2002) Potential Repair for *Escherichia coli* DNA Following Exposure to UV Radiation from Both Medium- and Low-Pressure UV Sources Used in Drinking Water Treatment. *Appl. Environ. Microbiol.*, **68** (7), 3293–3299.

9.0   SUGGESTED READINGS

American Society of Civil Engineers; American Water Works Association (1990) *Water Treatment Plant Design*; McGraw-Hill: New York.

Arizona State University; University of Arizona; University of Colorado at Boulder; U.S. Water Conservation Labor; Greeley and Hansen (1996) *Soil Treatability Pilot Studies to Design and Model a Soil Aquifer Treatment System*; Draft Short Report; Arizona State University: Tempe, Arizona.

Asano, T. (1985) *Artificial Recharge of Groundwater*; Butterworth Publishers: Stoneham, Massachusetts.

California Department of Health Services (2007) *Groundwater Recharge Draft Regulations*; Sacramento, California.

Crook, J. (1990) Water Reclamation. In *Encyclopedia of Physical Science and Technology, 1990 Yearbook*; Academic Press: San Diego, California.

Drewes, J. E.; Quanrud, D.; Amy, G.; Westerhoff, P. (2006) Character of Organic Matter in Soil–Aquifer Treatment Systems. *J. Environ. Eng.*, **11**, 1447–1458.

U.S. Environmental Protection Agency (1988) *Constructed Wetlands and Aquatic Plant Systems for Municipal Wastewater Treatment*; EPA-625/1-88-022; Center for Environmental Research Information: U.S. Environmental Protection Agency: Cincinnati, Ohio.

Chapter 5

System Reliability

1.0   INTRODUCTION

Reliability is the bottom line in any potable water supply system. The guiding principle for water resource planners is to always choose the best source of water for the potable water supply. In this context, the term best depends on location, regional availability of water, and time. As water resources become scarce and energy costs increase, eyes will be focused on sources that may not be considered "best" today. A water source with little variation in quantity and quality that is independent of weather fluctuations deserves consideration. From the standpoint of quantity, reclaimed water is one of the most reliable sources of water available in sewered communities. The challenge for water reclamation agencies is to guarantee water-quality reliability as well. This is the focus of this chapter.

To ensure that reclaimed water is reliable in both quantity and quality, the treatment facilities must be properly designed. If breakthrough occurs in any process, the treatment system must be flexible enough to compensate without a loss in discharged water quality; to do this, multiple backups and barriers must exist. This chapter will examine the place of storage, filtration, disinfection, and soil–aquifer treatment (SAT) in a multibarrier design approach. (Although existing technologies can convert wastewater directly into potable water, this chapter focuses solely on indirect potable reuse.)

Once a treatment system is constructed and operating, sampling, monitoring, and permit compliance are keys to ensuring overall reliability. Another key is a procedure manual that succinctly addresses proper operations and immediate correction of potential upsets, if they occur. Standard operating procedures for a variety of processes will be examined. Operator training issues and emergency response procedures are also covered in this chapter. Regulators' role in assuring reliability—from planning through design and operation—is also addressed.

This chapter will provide insight into the issues that help ensure that a water reclamation facility reliably produces the water quantity and quality needed to augment potable water supplies. Such reliability is the key to responding to public concerns and perceptions.

2.0   RELIABLE DESIGN AND OPERATIONS

2.1   Treatment Facilities

Water reclamation facilities must be designed to provide maximum reliability at all times; those involved in indirect potable reuse must have features to prevent the conveyance and use of reclaimed water that may be inadequately treated because of a process upset, power outage, or equipment failure. (For more information on treatment process technologies, see Chapter 4.)

The fundamental goal of treatment system reliability is to provide reclaimed water that meets or exceeds water-quality and -quantity standards. An array of design features and provisions may be used to improve the reliability of both process components and the system as a whole. Because state criteria vary and each water reuse project is unique, each project must be tailored to meet, at a minimum, the specific reliability requirements determined by regulators. Such requirements include

- Alarms,
- Standby power supplies,
- Readily available replacement equipment,
- Multiple or standby treatment units,
- Emergency storage and disposal or re-treatment provisions for inadequately treated wastewater,
- Piping and pumping flexibility to permit flow rerouting during emergencies, and
- No options to bypass any treatment processes.

A conservative design approach is necessary to preserve multiple-barrier protection and adhere to regulatory guidelines despite process upsets and other unforeseen conditions. However, no design, no matter how conservative, will provide adequate reliability without enough staff and properly trained, certified operators.

2.1.1 Process and Equipment Redundancy

When treatment process units are taken out of service for maintenance or repair, redundant or standby units should be available to continue treatment. Otherwise, facilities need to be able to divert the wastewater to other approved disposal facilities. Similarly, if vital components [e.g., the filtration system, disinfection system, and power supply (WPCF, 1989)] fail, a system of backups must be in place to take over critical functions. Within treatment trains, facilities depend on downstream processes to help absorb upstream process upsets or mechanical failures. Enough redundancy should be provided to prevent any process or mechanical component from becoming absolutely vital to the protection of public health.

The U.S. Environmental Protection Agency (U.S. EPA) has developed water reclamation plant reliability classes based on the receiving water's use and the probable adverse effect of inadequately treated discharge. Because of the public health considerations associated with indirect potable reuse, U.S. EPA's most stringent category of reliability requirements (Class I) should be considered a minimum standard for indirect potable reuse applications, unless state regulations are more restrictive or states allow alternate methods of achieving reliability. Class I reliability requires redundant units to prevent treatment upsets during power and equipment failures, flooding, peak loads, and maintenance

shutdowns (Table 5.1). Some states may have more stringent reliability requirements that depend on the plant's rated capacity (WEF, 1992); others (e.g.,
Florida) allow alternatives to Class I reliability (e.g., providing another disposal
option with less-stringent treatment standards when indirect potable reuse requirements are not met). California has stringent reliability requirements for
the chemical coagulation, flocculation, and disinfection processes, as well as
process monitoring and control (California Code of Regulations, Title 22, Div 4,
Chapter 3, §60333–60355) For more information, see the California Department
of Public Health, Division of Drinking Water and Environmental Management
Web site: www.cdph.ca.gov/healthinfo/environhealth/water/Pages/Water
recycling.aspx.

TABLE 5.1   Summary of Class I reliability requirements (U.S. EPA, 1992).

Unit	Class I requirement
Mechanically cleaned bar screen	A backup bar screen shall be provided (may be manually cleaned)
Pumps	A backup pump shall be provided for each set of pumps that performs the same function; design flow will be maintained when the largest unit is out of service
Comminution facilities	If comminution is provided, an overflow bypass with bar screen shall be provided
Primary sedimentation basins	There shall be sufficient capacity to maintain 50% of total design flow capacity when the largest unit is out of service
Filters	There shall be a sufficient number of units, which are sized to maintain at least 75% of the total design flow capacity when one unit is out of service
Aeration basins	At least two basins of equal volume shall be provided
Mechanical aerator	At least two mechanical aerators shall be provided; design oxygen transfer will be maintained when one unit is out of service
Chemical flash mixer	At least two basins or a backup means of mixing chemicals separate from basins shall be provided
Final sedimentation basins	There shall be a sufficient number of units, which are sized to maintain 75% of the design capacity when the largest unit is out of service
Flocculation basins	At least two basins shall be provided
Disinfectant contact basins	There shall be a sufficient number of units, which are sized to treat 50% of the total design flow when the largest unit is out of service

2.1.2　*Piping and Pumping Flexibility*

The process piping, equipment arrangement, and unit structures must allow for operations and maintenance (O&M) ease and efficiency, as well as maximum operating flexibility. Flexibility will permit the necessary degree of treatment to be obtained under varying conditions. It can take the form of piping to alternative locations for chemical coagulant or flocculant addition, piping to recycle inadequately treated water back to the plant inlet, etc. Also, all aspects of plant design should allow for routine maintenance of treatment units without diminishing water quality.

No pipes or pumps should be installed that would allow bypass of critical treatment processes or permit inadequately treated effluent to enter the reclaimed-water transmission system. The facility should be able to operate during power failures, peak loads, equipment failure, treatment plant upsets, and maintenance shutdowns. In some cases, it may be necessary to divert the reclaimed water to emergency storage facilities or discharge it to approved, less-restrictive or nonreuse areas. During periods of power or equipment failure, standby portable diesel engine-driven pumps also can be used (U.S. EPA, 1992).

2.1.3　*Disinfection*

From a public health standpoint, an adequate, reliable disinfection process is the treatment system's most essential feature (WPCF, 1989). Chlorination, the most widely used disinfection process, can be interrupted by various causes (e.g., exhaustion of chlorine supply, chlorinator failure, water supply failure, and, most commonly, power failure). The following features can make chlorine disinfection processes more reliable:

- Standby hypochlorite storage,
- Standby chlorine cylinders,
- Chlorine cylinder scales and tank-level indicators,
- Manifold systems (to ensure that standby cylinders can take over for exhausted ones),
- Alarm systems (to notify staff about loss of chlorine, ejector water supply, or hypochlorite feed),
- Standby chlorinators and hypochlorite pumps,
- Automatic cylinder changeover or switch to the standby metering pump,
- Multiple-point chlorination,
- Automatic control of chlorine dose, and
- Automatic measuring and recording of chlorine residual.

Spare cylinders should be available if continuous chlorination will be provided. Scales help identify the amount of chlorine remaining in a cylinder so the need to replace it can be anticipated. A manifold system allows for rapid changeovers to full cylinders, as well as a greater chlorine reserve and longer intervals between cylinder changes. Automatic cylinder-changeover devices on the manifold system can provide uninterrupted chlorination without operator attention (U.S. EPA, 1992). Because of stringent U.S. EPA Risk Management Program requirements and some state requirements for chlorine gas above threshold quantities, liquid chlorine (hypochlorite) is becoming more prevalent among utilities using chlorine disinfection. Also, some states may mandate dechlorination, which should be factored into the choice of disinfectant.

Many systems may use alternative disinfectants [e.g., ozone or UV irradiation]. To the extent possible, such systems' reliability and redundancy requirements should be equivalent to those for chlorination. For ozone systems, critical considerations include adequate oxygen supply, redundant ozone-generation equipment and contact chambers, on-line monitoring of ozone feed gas and ozone residual, automatic adjustment of feed rates, alarms, and automatic system shutdown or effluent diversion if ozone supply is lost. For UV systems, critical considerations include redundant contact chambers, velocity distribution through the UV channel, adequate supply of spare lamps, on-line monitoring of transmittance, UV intensity, and automatic shutdown or diversion if the lamps lose their power supply (NWRI, 2003).

In some systems, physical separation techniques (e.g., membrane filtration) are a critical element of the disinfection process, so similar consideration should be given to these systems in terms of equipment redundancy, on-line monitoring of performance, and overall membrane and treatment integrity. Potable water treatment facilities using membrane filtration must perform integrity testing, so water reclamation systems should do the same.

2.2   Emergency Storage or Discharge

Reclamation systems that augment potable supplies should provide a reliable means of diverting reclaimed water away from the potable-water storage and delivery system when it does not meet regulatory requirements or the project's water-quality goals. An emergency storage facility or alternate discharge arrangement is an essential element of a comprehensive reliability program.

When selecting the most appropriate storage or discharge method, the project team should conduct a site-specific analysis of available options. The foremost consideration should be each option's reliability, so those directly under the reclamation agency's control are attractive. Potential emergency management options include the following:

- Discharge to a receiving water (if the reclaimed water meets water-quality requirements);
- Diversion to a less-demanding reuse application (e.g., irrigation) (if the water quality is acceptable);

- Discharge to another reclamation facility's collection system;
- Discharge via approved methods that do not contribute to potable water supply (e.g., deep-well injection, evaporation or percolation ponds, and spray disposal areas);
- Diversion to holding ponds, tanks, or any other facility reserved for emergency storage or discharge.

To the extent possible, diversion to emergency storage or discharge facilities must be automatically controlled via in-line (real-time) sensors monitoring critical water-quality parameters. This requires reliable instrumentation and mechanical systems (e.g., continuous-recording turbidimeters, particle counters, transmittance, and disinfectant-residual sensors) and ongoing maintenance and calibration of primary sensors. Testing may be warranted to develop surrogate parameters [e.g., total organic carbon (TOC)] that may lend themselves to in-line monitoring.

Reliable in-line monitoring equipment is not available for all water-quality parameters, so the project team should establish a manual diversion protocol based on a routine sampling program. The diversion system should include a manual reset feature to prevent automatic delivery of unsuitable water until the excursion is corrected and reclaimed water is back within specified limits. All emergency-diversion system components, except return pumping equipment from holding ponds, should have a standby power source.

To determine the water-quality conditions that would trigger an automatic or manual diversion of reclaimed water, the project team should consult with the appropriate regulators. An example of such a program was the monitoring and diversion protocol for the proposed San Diego Water Repurification Project (Bernados, 1996) (see Figure 5.1 and Table 5.2). Implementation of this Repurification Project has been deferred indefinitely; however, the information in Table 5.2 may be applicable elsewhere. The monitoring parameters and the alarm levels are only suggestions and may not necessarily be appropriate for all projects in other locations. Furthermore, should this Repurification Project be reconsidered, the California Department of Public Health (formerly Department of Health Services) will review the new proposal and make appropriate modifications to the monitored parameters and alarm levels and may even consider monitoring other parameters as regulations and technology warrant. In the original Repurification Project, online (real-time) monitoring was to be used to detect treatment failures and initiate diversions. Monitoring for appropriate parameters would follow each significant unit process in the multibarrier treatment train as shown in Figure 5.1. If a breakthrough were detected, an alarm would be triggered and the inadequately treated water would be diverted to a sewer for conveyance to the Point Loma Wastewater Treatment Facility downstream. Table 5.2 lists the monitoring strategies and water quality conditions that were

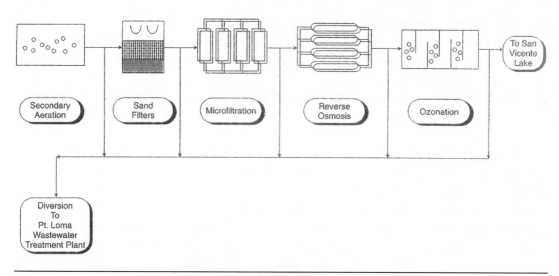

FIGURE 5.1 Diversion points for inadequately treated water.

proposed for the Repurification Project that would trigger an alarm and diversion. Online monitoring was to be supplemented by additional direct testing procedures to ensure treatment reliability and integrity.

Note that in Table 5.2, turbidimeters are listed as monitoring devices for water quality. The state of Florida has allowed a low solids TSS meter in lieu of a turbidimeter to monitor and control the filtration process in a high level disinfection facility. The device may offer advantages over a turbidimeter.

When an emergency storage facility is a reliability feature, its holding capacity should be able to accommodate flows throughout the treatment-disruption

TABLE 5.2 Proposed real-time monitoring and alarms for the proposed San Diego Water Repurification Project (Bernados, 1996).

Process	Monitoring device	Unit	Alarm
Secondary	Turbidimeters	NTU	10
Tertiary	Turbidimeters	NTU	5
Microfiltration	Turbidimeters	NTU	0.1
	Particle counters	Number in a size range	Abnormal increase
	Pressure transducers	Transmembrane psi	Abnormal increase
	TOC analyzer	TOC	15
Reverse osmosis	Particle counters	Number in a size range	Abnormal increase
	Pressure transducers	Transmembrane psi	Abnormal increase
	TOC analyzer	TOC	5*
	Flow meter	Transmembrane flux	Abnormal increase
	Conductivity meter	$\mu.mho/cm$	Abnormal increase
Ozone	Residual analyzers	mg/L	No residual
	Feed gas monitor	% per weight	0

*It is possible that this limit could be proposed at a level significantly lower; perhaps as low as 0.5 mg/L or lower.

period. In addition, the treatment facility should have enough capacity to accommodate both normal influent flow and the return flow from storage after the diversion event. Short-term holding tanks whose retention time can meet the longest experienced or anticipated power outage—plus additional reserve—should be adequate for problems caused by interruptable power supplies. (Providing automatic standby power is not necessarily a substitute for emergency storage because the generator could fail to start or the power interruption could be caused by a faulty breaker, transformer, or power cable.) Such tanks also may be suitable in situations where replacement parts are immediately available and corrective actions can be performed quickly.

Unfortunately, treatment disruptions or equipment failures may not be short-term events, so facility designers should consider making a conservative estimate of the potential for treatment failure and the amount of system downtime required. One guideline used for nonpotable reuse applications is 20 days or more of emergency-storage retention time, based on the rationale that this is long enough to complete almost any necessary corrective action (WPCF, 1989). For potable reuse systems, the team should conduct a project-specific critical analysis of emergency storage needs.

2.3   Transmission and Conveyance

The most important reliability issue for indirect-reuse distribution systems is protecting water quality in the pipelines that convey reclaimed water to a potable surface water reservoir or groundwater-recharge site. Because these pipelines deliver supplemental water, ensuring that they are delivering water at all times is not an important consideration. (In a potable water distribution system, however, customers expect water at all times.)

To prevent accidental damage or an unintentional cross-connection, pipelines must be properly identified by a special covering or coating and labels in multiple languages. When crossing wastewater lines or lower-quality reclaimed water lines, the pipeline's clearances, materials, and protection must conform to local regulations. Air and vacuum valves must be protected from flooding (potential intrusion of contaminated water) and vandalism.

2.4   Reuse Facilities

In this chapter, *reuse facilities* include groundwater-recharge sites and surface water reservoirs.

2.4.1   Groundwater Recharge

There are two types of groundwater-recharge systems: surface spreading and direct injection. In surface spreading, reclaimed water is applied to open basins at the ground surface and percolated through the soil to the aquifer. In direct injection, reclaimed water is pumped or flows by gravity into the groundwater aquifer via a well. This technique is particularly effective when creating fresh-water barriers in coastal aquifers to prevent saltwater intrusion.

Groundwater-based potable water systems typically extract the water, disinfect it, and then distribute it. If the groundwater is augmented by reclaimed water via surface spreading, SAT (soil–aquifer treatment) is relied on to treat the reclaimed water as it moves from the recharge point to the withdrawal point. The soil–aquifer treatment process' effectiveness and reliability depend on the groundwater-recharge method and the physical, chemical, and hydrogeological characteristics of the aquifer recharge and withdrawal system. When direct injection is used, the benefit of SAT is minimized, so regulators typically require higher-quality reclaimed water for such projects.

2.4.2   Recharge System Reliability

At a minimum, the design, management, and operation of the groundwater recharge and extraction system must conform to local regulations. The potable water supplier may have more requirements [e.g., depth to water, aquifer residence time, separation between the recharge area and the extraction well(s), and blending with nonreclaimed water]. Several states have extensive requirements (e.g., State of California, 2007). (For more information, see Chapter 3.)

2.4.3   Surface Water Replenishment

When reclaimed water is used to augment surface water supplies, the water reclamation facility's reliability and treatment requirements are often higher than those for facilities involved in groundwater-recharge operations because surface water does not provide the same residence-time and SAT benefits. Instead, they must rely on an effective water treatment plant to provide another barrier against contaminants. Such plants typically provide coagulation, sedimentation, filtration, and disinfection. Depending on water-quality conditions, they also may include more advanced processes to remove specific contaminants. More reliability and design "robustness" may be required for these water treatment plants than for conventional ones.

2.5   Instrumentation, Controls, and Alarms

Instrumentation and control systems for on-line monitoring of treatment process performance and alarms for process malfunctions merit special consideration. The key reliability considerations for each are highlighted below.

Considerations when developing instrumentation and control systems for reclamation facilities include the

- Ability to analyze appropriate parameters on-line reliably,
- Availability of qualified staff to calibrate and maintain the instrumentation,

- Identification of critical treatment processes that require monitoring and control,
- Identification of reclaimed water conveyance monitoring and control needs,
- Alarms needed to notify operating personnel, and
- Need for operating personnel onsite or on-call.

The potential uses of reclaimed water determine the degree of instrument sophistication required in a water reuse system. There are certain risks when wastewater is being treated for indirect potable reuse via groundwater recharge or surface water augmentation, so instruments must be highly reliable and sensitive enough to immediately detect even minor discrepancies in water quality (U.S. EPA, 1992). Instruments that require frequent calibration or have questionable reliability will often be bypassed to avoid "nuisance" alarms (which defeat their purpose).

Alarm systems are installed at water reclamation plants to effectively warn of and minimize interruptions in treatment processes. Alarms and annunciators monitor the condition of equipment. Equipment failure may cause the following conditions (WPCF, 1989):

- Damage to a vital process or piece of mechanical equipment;
- Damage to a component that would be costly, difficult, or time-consuming to replace, whether it is part of a critical or noncritical process;
- A hazard to operating personnel; or
- Compromised product quality.

Minimum instrumentation consists of alarms at critical treatment units to alert an operator of a malfunction. Often, regulators will have specific alarm and monitoring requirements. Warning-system requirements should specify the measurement to be used as the setpoint in determining a unit failure (e.g., transmittance in UV irradiation influent, or turbidity in the chlorination influent). Alarms may be actuated by various situations (e.g., power failure, high water level, pump or blower failure, loss of chlorine residual, loss of coagulant feed, high filter headloss, high effluent turbidity, or loss of chlorine supply). Many supervisory control and data acquisition (SCADA) systems allow operating personnel access to the system via remote "laptop" computers, which can save time and resources.

There should be provisions for treating, storing, or disposing of wastewater until corrective actions have been made.

2.6   Power Supply

Electrical power systems also merit special consideration. Water reclamation plants should have a standby power source for all critical process components, including key monitoring and emergency-storage diversion or discharge equipment. This source should have enough capacity to provide necessary service during normal power-supply failures. Standby power sources typically include gasoline- or diesel-operated generators, or connections to another power grid. Each power source should have its own transformer and motor control center. Many reclamation plants provide standby power via engine-driven generators that require manual starting; they would be more reliable with battery-operated switchover mechanisms and automatic starters. Standard operating procedures should require testing of all equipment, under full or partial load, at least once a week.

It may be necessary for the primary power source to sustain only critical loads in a standby or emergency mode of operation. Critical loads must include the instrumentation and control system, critical lighting and ventilation, and the systems for diverting recycled water to emergency storage or less-critical reuse applications. Instrumentation and control systems should have a battery backup (an uninterruptible power supply) to take over regardless of auxiliary power. Depending on whether treatment must continue during power outages, other equipment (filtration and disinfection processes) will need to be on auxiliary power. One source of electrical power typically is sufficient to handle the needs of noncritical operations.

Power distributed to the plant's main control centers or control panels for critical loads typically is supplied by motor control centers connected to in-plant unit substations. There should be redundant substations and feeders to motor control centers. Critical in-plant power loads should be divided in the motor control center by tie breakers. The motor control center should be supplied with power at all times to treat the reclaimed water. The instrumentation and control panels associated with the operation of critical process loads also should have similar redundancy (U.S. EPA, 1992).

2.7   Operations

Potable water treatment is based on the principle of establishing multiple barriers to prevent—to the greatest extent practical—pathogens and harmful organic and inorganic contaminants from entering the drinking water system. Multiple barriers are important because one barrier may not be able to accomplish all tasks, and multiple barriers provide backup if any one barrier fails or partially fails.

Multiple barriers also are required for water reclamation facilities that are augmenting potable water supplies. Such barriers include secondary and tertiary treatment, disinfection, monitoring, and effective operating procedures. They also include SAT or surface water treatment facilities.

This section describes the operating procedures, training, and response actions needed to ensure the quality of reclaimed water. The emphasis is on the

actions, tests, and procedures that are different from or in addition to those typically performed at a water reclamation plant that provides flow for irrigation
and other nonpotable uses.

2.7.1 Standard Operating Procedures

The standard operating procedures for a water reclamation plant that provides
for nonpotable reuse (e.g., irrigation or industrial processes) typically include
"minimum" process monitoring to ensure conformance to permit requirements.
(This is not to say that all agencies provide only minimum process monitoring;
those with extensive resources frequently go beyond the minimum.) The standard operating procedures for water reclamation facilities that provide for indirect potable reuse go beyond the permit minimum—especially if the water is
discharged into a potable water reservoir or tributary.

A water reclamation facility's operating and monitoring procedures typically are identified in its O&M manual. If the facility augments a potable water
supply, its O&M manual should be more explicit in terms of monitoring and
controlling pathogens and carcinogens that may have some long-term effect. The
manual also should include a copy of the approval, permit, or other license so
operators can reference it and become conversant with the permit limits.

Standard operating and monitoring procedures must be detailed enough to
ensure that the barriers keeping carcinogens out of the water supply are not penetrated. Both the O&M manual and associated training should include a discussion of the characteristics of various types of pathogens and carcinogens, as well
as how they can be kept out of both the wastewater and reclaimed water. The
discussion should include both physical (filters, membrane and UV systems,
etc.) and chemical (disinfection) barriers. It should focus on each barrier's function, operations, and limitations, and note how it can be penetrated.

Also, an emergency response plan is needed to ensure seamless communication between responders and the reclamation plant if toxics or other hazardous
chemicals are spilled into the wastewater collection system.

Chapter 3 discusses the health implications of using reclaimed water. Of importance from an operations standpoint is the presence of pathogenic protozoa
(e.g., *Giardia lamblia* cysts and *Cryptosporidium parvum* oocysts), which are resistant to chlorine disinfection. These pathogens are a concern in nonpotable reuse,
but even more so when the reclaimed water is used to augment potable water resources, where the exposure via ingestion is substantially greater.

Fortunately, a properly designed and operated sand or anthracite media
filter effectively removes *Giardia* and *Cryptosporidium* from water. Although protozoa are larger than bacteria and viruses, they are small enough to penetrate
an improperly operated filter (*Giardia lamblia* cysts range from 5 to 15 μm, and
Cryptosporidium parvum oocysts range from 4 to 7 μm). The March 1993 cryptosporidiosis outbreak in Milwaukee attests to this problem. Monitoring the

filtered water's turbidity (the traditional surrogate parameter for pathogen removal) is extremely important, but it may not be enough in some cases. On-line particle counters may monitor filtration results more effectively. Researchers found lower correlations between turbidity and *Giardia* and *Cryptosporidium* removals than between particle counts and *Giardia* and *Cryptosporidium* removals. Particle removal in the 4- to 7-μm range correlated reasonably with *Cryptosporidium* removal, and particle removal in the 7- to 11-μm range correlated significantly with *Giardia* removal (Nieminski and Ongerth, 1995).

If a water reclamation facility uses membrane filtration, the units also must be monitored for breakthrough and membrane integrity. This includes on-line monitoring of flux rate, transmembrane pressure, filtrate flowrate, and so on to evaluate both the process and membrane life (inorganic fouling, organic fouling, membrane life, cleaning periods, etc). Particle counters also might be effective.

Monitoring reclaimed water from each filter or membrane filter bank is essential to ensure that media or membrane failure or excessive headloss, which will cause breakthrough of particles (and presumably *Giardia* and *Cryptosporidium*), do not occur. It is important that operators be trained in proper filter operations, including backwash and potential filter-to-waste, if required. Proper operation is more critical in facilities involved in indirect potable reuse than in those involved in nonpotable reuse. The disinfection system—the last barrier before water leaves the reclamation plant—must be carefully monitored to ensure that adequate disinfection is occurring at all times.

Fortunately, reuse projects involving surface spreading or direct injection into groundwater have additional barriers in the form of SAT, which provides subsurface microorganism filtration and underground residence time. Undoubtedly, subsurface filtration is more effective than conventional filtration processes because of the substantially reduced subsurface flow rates and nonuniform particle sizes, which translate into reduced pore size. Over time, subsurface filtration ensures a nearly pathogen-free well water supply. However, caution should be taken if fractured rock aquifers, which can provide a direct conduit to remote wells, are present. It is important to thoroughly characterize the subsurface conditions and groundwater quality before initiating design.

To avoid introducing inorganic and organic contaminants to the water supply via reclaimed water, such contaminants should be controlled at their sources via a stringent, aggressively enforced industrial waste ordinance. Public education is also essential because a number of these contaminants originate in households.

Reclaimed water used to augment potable supplies should fully comply with established federal and state drinking water standards before it leaves the treatment facility. Staff also should routinely monitor trace elements and contaminants in reclaimed water for which regulations may be pending. It is important to identify and quantify as many trace organics in recycled water as possible to minimize the "unknowns" in the recycled water TOC. These data can be used to form a baseline and identify trends, so corrective action may be taken if needed. The data also may be used to form a baseline to monitor the fate of these constituents once they enter the SAT system or a tributary to a potable water reservoir.

In addition, it is important that a recharge site be designed to permit effective monitoring. This includes multiport, multi-level wells (or equivalent) that sample at specific depths below the surface. The fate of microorganisms, trace elements, and trace organics in the SAT system should be monitored routinely and under different operating scenarios (e.g., wetting and drying cycles and resting cycles) to optimize the removal of trace contaminants and organics, microorganisms, and inorganic constituents (e.g., nitrogen).

2.7.2   Cross-Training

Wastewater reclamation plant operations often are subdivided into departments or divisions—especially at larger facilities. For example, one division or supervisor may operate the primary and secondary systems, and another may operate the tertiary system. Because of the effect that changes in one part of a reclamation plant can have on other processes, standard communication procedures should be developed and implemented. For example, a process change in a secondary treatment system may temporarily increase suspended solids, which will affect tertiary system performance. A change in filter operations may unexpectedly return excessive amounts of filter backwash to the primary and secondary systems, causing a temporary hydraulic overload that may affect performance. So, pending and immediate changes to any process, including routine maintenance shutdowns, must be communicated to each department. With proper warning, recycled water can be diverted from the recharge area if necessary.

Operations personnel should be cross-trained in each facility's divisions and procedures. In addition to providing backup personnel, which is particularly important at smaller facilities, cross-training provides an awareness of other divisions' concerns. Awareness is the first and most important step toward effective communication.

Communication becomes more complex when multiple agencies are involved. Often, the water reclamation and potable water agencies are different entities. When the agencies are remotely involved (e.g., a water purveyor with wells near the recharge site or that operates the downstream potable water reservoir and treatment plant), communication must overcome agency hierarchy and protocol. Lines of communication should be established before using reclaimed water and regularly monitored for effectiveness thereafter.

If the potable-water supply agency is operating the reclamation facilities on the same site as the wastewater treatment facilities, the training and communication procedures are similar to those for one agency with multiple departments.

However, each agency's hierarchy and protocol may be a barrier to effective, rapid communication, and should be addressed as such.

Cross-training each agency's operators is desirable to improve communications and emergency response. However, using each agency's operators as a routine backup for each other is likely to be administratively awkward, so it should not be the primary reason for cross-training.

2.7.3 Emergency Response

Any number of emergencies may occur at a water reclamation facility—a treatment system upset, a traffic accident or chemical spill that inadvertently releases toxic materials into the wastewater collection system, or a treated-water pipeline break that might allow contamination to enter the pipeline and, eventually, the reuse site (e.g., a groundwater injection well). Also, depending on the facility's permitting requirements, a pipeline break that causes reclaimed water to be released from the transmission system or place of use may be a permit violation requiring rapid response. Effective, rapid responses to such emergencies depend on careful planning.

One of the first steps is developing a mutual aid agreement with other local agencies. This agreement sets forth the terms and procedures for securing personnel, equipment, and other resources from one another. It should identify each agency's emergency response coordinator and outline various means of communication. Many agencies develop emergency communication systems (e.g., radio links) with other local agencies in case of telephone outages or poor cell-phone signal strength. Having trained amateur radio operators (hams) on staff provides more emergency communications capability (Boyle Engineering Corporation, 1990).

It is often useful to issue special identification cards or badges to response personnel that are recognizable to law enforcement and other agencies that may control the ingress and egress of people at a disaster site. Unless an individual is recognized as an agency's emergency respondent, he or she might not be allowed access to the disaster site. If personnel may have to respond to emergencies in their own vehicles, they should have magnetic agency logos that can be quickly placed on vehicle doors to facilitate access.

In addition to identifying external lines of communication, internal lines of communication should be established (i.e., who must be notified and in what order, who is on call, etc.). Having a contract in place with equipment-rental firms and local construction contractors can facilitate rapid response. Checklists should be developed and priorities established for inspecting and responding to an emergency. Event-specific plans also should be developed for any reasonably anticipated emergency, because all emergencies do not require the same type of response. Such plans should detail the actions to be taken, the equipment and materials to be used, etc.

Without training, even the best plans are useless. Training should be done frequently for in-house emergencies and periodically for multiagency emergencies. Facilities that directly augment potable water supplies should have personnel cross-trained. Wastewater treatment plant operators should be trained to respond to emergencies in a reclaimed water system, just like distribution system operators respond to potable water system emergencies.

3.0   ENSURING WATER QUALITY

Water-quality management encompasses several topics a designer should consider to ensure that a potable reuse system operates successfully. The following water-quality management topics have unique requirements or characteristics for potable reuse. This section is not intended to be a complete guide to preparing a water-quality management plan; rather, it is intended to help a project team apply the existing body of knowledge on this subject to tailor a water-quality management plan for potable reuse. (For more information on designing and preparing a plan, see the many existing references on managing water quality.)

3.1   Regulatory and Performance Compliance

The goal of any operating system is to maintain compliance with a project's standards and requirements. While the most familiar compliance standards are numerical constituent requirements established by the appropriate regulatory agency, a successfully operated potable reuse system has established performance measures to ensure compliance not only with regulatory requirements but also with the design, political, managerial, or public expectations for the project. Performance compliance is essentially a measuring system that compares operating system performance to expectations, documents compliance, and identifies performance variations for early evaluation and correction.

The primary purpose of performance compliance is to ensure that the operating system efficiently accomplishes the water quality goals established during the design phase. Performance criteria can be developed based on the project's design criteria and key operating characteristics. Each treatment process should have at least one performance measure, but the number of routinely used performance measures should be the minimum necessary to ensure that the system is operating as desired while minimizing costs. Measurement frequency should also be the minimum required to ensure performance compliance. *Performance compliance* is a system management tool to ensure that the process is performing as expected; it is not intended to provide all the information necessary for a complete process analysis if a violation occurs. If a deviation develops, then more investigation may be required to pinpoint the problem.

Potable reuse systems may require new, unfamiliar, or unique performance measures. Consider, for example, a groundwater-injection project in which reverse osmosis is used to remove TOC and provide a virus barrier. The performance measure may be electrical conductivity (reverse osmosis also removes salts when operating properly, so electrical conductivity is an instantaneous measurement that may be monitored remotely). A change in electrical conductivity immediately indicates a reverse osmosis failure and, by inference, a failure in TOC removal and possibly the pathogen barrier.

At the point where reclaimed water is introduced to the water supply, the performance measure may be a tracer element (e.g., fluoride) which may be used to ensure that the water is mixing properly, not short-circuiting.

An important aspect of performance compliance is establishing and maintaining a database, which provides the necessary background information to monitor long-term trends and helps identify developing impacts. Even a simple analysis comparing data over time will provide a basis for assessing the suitability of performance measures. A good data record will demonstrate reliability and help establish public confidence in system operations.

Regulatory permit requirements are the most familiar source of performance measures. Most of these include both numerical performance requirement and measurement frequency. Treatment processes or techniques also may be specified. While most regulatory requirements are established before a system is started up, they may be modified as the system develops an operating history. This is particularly true for indirect potable reuse systems because the concept is new to most regulators and a set of permit requirements has not been established. It is essential that an operator or manager understand why each constituent is important to regulators and the significance of each performance measure, so he or she can provide meaningful input for modifying the requirement, as appropriate.

The final sources of performance compliance criteria—political, managerial, and public—are unique to potable reuse systems. Potable reuse projects typically involve a high level of public involvement because they often are the first of their kind in an area. During the development process, specific promises may be made about the system's ultimate performance. Successful potable reuse systems recognize such promises, measure them once the system is in operation, and communicate the monitoring results. Political, managerial, and public performance criteria are not always reflected in specific design parameters, so the public record is often a source of essential performance compliance criteria to ensure success.

Political, managerial, and public performance requirements often can be translated directly into measurable requirements. In a groundwater injection project, for example, the regulatory requirement may stipulate that no more than a certain percentage of the water extracted for potable use can be of wastewater origin. The public may be concerned that an underground buildup is occurring and that violations and permanent groundwater contamination are inevitable. While the regulatory requirement can be demonstrated with extraction wellhead testing, the performance measure may be a series of sampling wells to monitor the flow and blend percentage of injected water. These data can be used periodically to update a map showing the changes that are occurring. The map can be distributed to the public to demonstrate that the system will operate in the long term, as predicted and promised in public meetings.

In addition to being sensitive to public perception and concerns, an operator or manager often has to be resourceful to ensure successful performance compliance. An example of a proactive performance-compliance criterion is offering a quarterly open house at the facility. In other words, every 3 months a system operator opens the facility to the public and the media, provides a short presentation on the previous quarter's performance, and then conducts a tour of the potable reuse system. The quarterly update and tour are intended to show that the system is being operated and maintained adequately. This performance-compliance criterion has been used successfully to integrate an environmentally responsible project into the community.

Once a performance-compliance program has been established, monitoring is necessary to measure compliance. Monitoring is accomplished via sampling, which, for most potable reuse systems, has to occur in two environments: in the treatment facility and in the field. Sampling in the treatment facility occurs in the treatment processes or closed conduits that are under the control of the system operator. Such sampling is similar to that in any water or wastewater system, and any related sampling references are appropriate for potable reuse systems.

The sampling environment unique to indirect potable reuse systems is field sampling, which is unique in several ways. Some indirect potable reuse systems discharge reclaimed water in areas with a high level of public access (e.g., a spreading basin with multiple uses). This may lead to vandalism of sampling locations or cross-contamination. Field sampling also may be in areas operated jointly by two agencies; this may lead to a situation in which the sampling can be a joint effort of the two parties, thereby minimizing or sharing costs.

The third situation unique to indirect potable reuse is that regulators may require a system operator or manager to monitor another agency's compliance. Consider, for example, an indirect potable reuse project that involves directly injecting reclaimed water into a groundwater basin used as a drinking water source. The operator of the wells extracting water from the basin may have to take samples and make operational decisions to maintain an acceptable percentage of recycled water in the potable water supply. However, regulators may hold the operator of the reclaimed water system responsible for maintaining an acceptable blend ratio.

These factors need to be considered not only in the design and installation of sampling locations, but also in the operating protocols for such locations. Security, simplicity, and minimization are key considerations when developing sampling locations and procedures. The sampling program also may be subject to more direct public observation and scrutiny than typical water or wastewater sampling programs, so it is important to have a well-maintained sampling location, written sampling procedures, and trained staff who can respond directly to public inquiries while taking samples.

A good quality assurance program is a standard part of water-quality management for any water or wastewater treatment system. What is unique in indirect potable reuse is the high potential for public scrutiny and involvement. A quality assurance program can become an important and powerful public relations tool if it is written so the public can easily understand it. While most quality assurance plans are prepared internally, the project team should consider having third-party experts prepare it, especially if public health and safety have been concerns during planning. A "plain English" quality assurance plan helps give the public the message that the system will be operated safely, openly, honestly, and forthrightly.

Records for an indirect potable reuse system also may be subject to more public scrutiny than those for a water or wastewater treatment system. The records system must be organized and orderly, so the public can be given immediate access if anyone makes an inquiry. Any well-meaning delays to make data presentable to the public may be perceived as "trying to hide information", particularly if the reuse project was implemented amid a high level of public concern. Immediate access to operating records may build public confidence as operations commence.

3.2   Performance Contingencies

Unless a significant violation occurs, performance excursions in water and wastewater treatment systems are primarily the concern of system staff and regulators. At indirect potable reuse systems, however, public concern may be considerable over even minor excursions. To maintain public confidence in an indirect potable reuse system, special consideration must be given to how violations are defined, isolated, and corrected.

In this context, a *violation* is a treatment system measurement that is outside an established range or more than an established limit. The word does not unduly worry system operators, managers, and regulators because they are familiar with the process and its corrective procedures. The public, on the other hand, does not know this process and already may have concerns about indirect potable reuse. So, the public and the media may interpret a violation as an immediate, significant threat to health and welfare, even though it is not. An indirect potable reuse system's water-quality management plan should include a definition of "violation of performance compliance" and, in some cases, the process for correcting it. This material must be presented "in plain English" so the public can easily read and understand it.

Definitions of operating and regulatory standards are typically simple and straightforward. An operating standard is used to control a process; the regulatory or compliance standard is the legal performance standard. Operating standards are early warnings; they alert facility personnel that attention is needed to ensure that a treatment process is maintained within specified norms. Regulatory standards are more critical, but they may cause less public concern than a violation of political, management, or public standards. For regulatory standards, it is useful to maintain an information sheet describing the effect of a persistently higher level of a constituent. Such sheets provide a simple, quick, and credible way to give the public a frame of reference about the effect of a regulatory violation.

Political, management, and public standards are the most critical because they are developed, typically during the project development phase, to alleviate specific public concerns that already exist in the community. A violation may be an immediate confirmation to the public that its prior concerns were indeed correct. When these standards are established, a definition of violation should be prepared that includes a clear description of what the violation means, how the public and the media will be notified, and the procedure to correct it. One example of this type of violation is an increase in specific conductivity from a reverse osmosis unit, which has been presented to the public as assurance that the reclaimed water cannot contain viruses.

Violations are always a high priority, but in an indirect potable reuse system, violations will cause more public concern than those in a water or wastewater treatment system, especially if they involve political, management, or public standards. So, it is important that the notification and corrective process for a violation include the public in some way, just as a regulatory agency is notified of a regulatory violation.

3.3 Contingency Planning

Most water and wastewater utilities have procedures manuals for their water-quality management programs that are important to contingency planning. These manuals, which typically are fairly technical, are specifically designed for treatment system operators and managers. For a potable reuse system, the manual is an important public document that provides the concerned public a written commitment to how its health and safety will be ensured.

The manual's purpose is to document the water-quality management plan, provide guidance to system personnel, and provide assurance to the public. So, it must be clear, simple, and easy to understand. Technical appendices can contain more specific information for operations and technical staff. The main sections of the manual should be reserved for communicating information to the public. For example, if a violation of the RO system's electrical conductivity occurs, a page or section can be transmitted quickly to concerned parties. This page or section would define the violation's effect, the role of secondary barriers, the corrective procedures to be followed, and the long-term effect if uncorrected.

The procedures manual may be an indirect potable reuse system's most important public relations document in terms of building public confidence. By presenting the manual in a user-friendly format, information can be provided quickly before misinformation begins to circulate. Timeliness, even when presenting bad news, will build public confidence faster than any other management tool.

4.0   REGULATOR PARTICIPATION

Indirect reuse of reclaimed water is a relatively new program, so there is a lack of long-term, scientifically documented data on which to base regulations. This puts regulators in a difficult position when determining what should be required and/or allowed to happen. If design and discharge criteria are unavailable, project-specific regulations may have to be individually developed. So, including regulators on the project team may prove beneficial.

4.1   Project Development Phase

To avoid unnecessary delays and expenses, it is essential to understand how current regulations affect process performance and system reliability, and to be aware of any proposed regulations that may affect the project. A project designer should investigate regulations as early as possible and contact the administering agency. An early dialogue with regulators also may save research time because they typically are up to date on existing and proposed regulations.

Based on the discussions, develop a clear, workable, and well-documented proposal that conforms to existing and proposed regulations. Project planners should address all applicable national, regional, state, and local regulations—especially all site-specific criteria. Moreover, the intent of regulations may be interpreted differently and their enforcement may vary by location. In addition, as times have changed, so too have some regulations. So, one should not assume that a previously accepted process is guaranteed to be usable.

Interestingly, indirect potable reuse regulations are not developed in most cases because proposals typically are required before regulators will even consider the new use.

When starting a project, the project team should contact a local or regional water-quality regulator. Because the agency responsible for regulating water quality may vary by location, it is important to identify this entity, which permits will be necessary, and who will issue them. In many cases, one permit that covers the entire project may be sufficient. However, approvals from other regulatory agencies may be required to get one blanket permit. In addition, these agencies' regulations may require that they issue separate permits for different periods or changing conditions. Besides water-quality regulators, some of the more common agencies that may become involved are the U.S. Army Corps of Engineers and the local departments of building and safety, fire, fish and wildlife, and transportation.

Another factor—political and public support—is covered in Chapter 6; that effort should be well-coordinated when working with regulators. A project with public and political support may get a better response from regulators. However, keep in mind that regulators are not in the business of defending the merits of each project. In addition, issuing a permit does not mean that the regulatoty agency will bear any legal responsibility for the project.

It is important to address situations in which regulations contradict each other. For example, while public health agencies may require a high degree of disinfection and residual disinfectant, agencies interested in protecting aquatic life may consider residual disinfectants harmful. So, contradictory regulations should be resolved before proceeding with the project.

To facilitate project development effectively, agreements on definitions, criteria, time schedules, and commitments should be in writing and signed by the proper authorities. Knowing ahead of time what will be required by all involved personnel may prevent later controversies. Whenever possible, construction, operations, and maintenance staff should be included at summary meetings.

For projects in which new technology is proposed, the burden of proof will be on the project team, so before approaching regulators, the relevant information should be prepared based on long-term, scientific data. Project proposals whose new technology requires modifications to existing or proposed regulations are often costly and time-consuming. Unfortunately, many regulators have limited resources, so they probably will not have personnel to help develop or test a project. Again, make regulatory contacts with the appropriate agencies as soon as possible.

4.2   Implementation Phase—Design and Construction

Regulators typically have not been invited to participate in a project during the design and construction phase, when changes may occur for various reasons. If feasible, regular design and construction meetings should be held to bring all involved parties together and keep them informed. Sometimes, sending the appropriate regulator written notification of a project change is sufficient, as long as no controversy is involved. However, if major changes to a project are anticipated, it may be beneficial to have the appropriate regulator attend regular design and construction meetings. Even if a regulatory agency does not have staff to regularly attend the meetings, it is still good policy to invite them in case there are any questions. Moreover, this may result in a quicker response time if project changes occur.

Another means of keeping regulators apprised of the project's progress is to send them a copy of final construction plans and specifications (sometimes, this may be mandatory). Another beneficial procedure is to send regulators copies of minutes from regular design and construction meetings for informational purposes. In addition, inviting them to participate in inspection tours or opening ceremonies can foster better networking.

An O&M manual should be prepared during the design and construction phases. These manuals are often essential references for operators and agencies because they typically contain all of the project's "as-built" data, design data, and technical specifications. Some regulators require that a technical engineering report be submitted as part of an operating permit. Because most pertinent data are available during the design and construction of a facility, it is an opportune time to collate them for staff training, records, and regulatory submissions, as required.

4.3   Initial Startup and Operation and Maintenance

Regulators are aware that even the best-designed and constructed facility or system will not meet expectations without proper O&M. Properly designed and constructed facilities with adequate capacity for peak flows and redundancy must be supported and maintained by thoroughly trained and equipped personnel. Assuming that a facility was constructed satisfactorily, the next step should be to test its performance. This typically is called the *startup period*, when the anticipated range of operating conditions is tested as much as possible. During this period, it may be necessary to negotiate temporary water-quality or other permitted discharge limits with the appropriate regulators. Diversion to nonpotable uses also may be necessary until performance criteria can be ensured. Provisions to accommodate this should be incorporated in the facility design.

This period also is used to provide hands-on training for O&M personnel. Before initial startup, staff should have the opportunity to become thoroughly familiar with the O&M manual and attend classroom training as needed.

Startup documentation is important because these data will be used later when equipment needs replacing (e.g., UV lamp and burn-in period, membranes and hydration period, etc.). In addition, startup and commissioning data are invaluable when troubleshooting processes or bringing new equipment on-line.

Based on the results of startup tests, it may be appropriate to fine-tune the sampling and monitoring requirements in the operating permits. Sometimes regulators may want research to be conducted. At minimum, sampling and monitoring activities should be reviewed and documented for future discussion. Results from startup tests offer practical information for modifying or fine-tuning O&M manuals. The scope and frequency of required permit activities should be reviewed for all short-term activities covered during the test period. Make sure the test period includes the operation of all emergency equipment and facilities and the complete documentation of results. Moreover, corrections and changes should be made before a contractor leaves the site or his retention is released.

If permit limits cannot be met during the startup period, an alternative plan should be worked out with regulators to allow the problem to be corrected before a discharge limit is violated. Alternative means of handling discharge (e.g., returning flow to the sewer, recycling flow back into the plant, diverting flow to an acceptable storage area, or post-discharge treatment) should be incorporated as part of the action plan for regulators to consider. With indirect reuse, there is an implied higher public health risk factor that should be kept in mind at all times. Problems should be acknowledged as soon as possible and a reasonable time allowed for corrections. An agency's credibility should not be sacrificed under any circumstances, because lost trust cannot be regained easily. So, all violations should be documented and reported immediately to the permitting agency. The last thing a regulator wants to learn from the news media is that one of its permittees is in violation.

Occasionally, there may be a situation in which it is impossible to meet the required discharge limit because of physical, biological, or other factors. Sometimes local regulators may determine that a discharge requirement is unsuitable for the situation. In many cases, however, local regulators have no choice in the matter because their job is to administer and enforce regulations. Consulting the agency with authority to change the legislation may be investigated as a means of providing relief.

Finally, a regulatory agency that works on a project may, in some cases, demand that operators implement a training program for all new employees, and that refresher training be provided for existing employees. In addition, regulators may require that adequate spare parts and tools to perform both preventive and corrective maintenance be available if a system failure occurs.

5.0   REFERENCES

Bernados, B. (1996) The Importance of Reliability in Potable Reuse. *Proceedings of the AWWA/WEF 1996 Water Reuse Conference*; San Diego, California; American Water Works Association: Denver, Colorado; Water Environment Federation: Alexandria, Virginia.

Boyle Engineering Corporation (1990) Emergency Preparedness Study for the San Gabriel Valley Municipal Water District, Azusa, California.

National Water Research Institute (NWRI) (2003) *Ultraviolet Disinfection Guidelines for Drinking Water and Water Reuse*, 2nd ed.; National Water Research Institute: Fountain Valley, California.

Nieminski, E. C.; Ongerth, J. E. (1995) Removing *Giardia* and *Cryptosporidium* by Conventional Treatment and Direct Filtration. *J. Am. Water Works Assoc.*, 87, 96.

State of California (2007) Chapter 3: Recycling Criteria. In *Draft Proposed Regulation: Title 22, California Code of Regulations, Division 4*; Environmental Health: Sacramento, California.

U.S. Environmental Protection Agency (1992) *Guidelines for Water Reuse*; EPA-625/R-92-004; Center for Environmental Research Information: Cincinnati, Ohio.

Water Environment Federation (WEF) (1992) *Design of Municipal Wastewater Treatment Plants*; Manual of Practice No. 8; Water Environment Federation: Alexandria, Virginia.

Water Pollution Control Federation (WEF) (1989) *Water Reuse*; Manual of Practice No. SM-3; Water Pollution Control Federation: Alexandria, Virginia.

Chapter 6

Addressing Public Perceptions

1.0 INTRODUCTION

Much of the information in this chapter is based on work conducted under the National WateReuse Foundation (Alexandria, Virginia). This work includes a Phase 1 Report (Ruetten, 2004), which presents 25 best practices and detailed case study analyses, and a Web site that provides multiple tools for addressing many of the key issues described in this chapter (see www.watereuse.org). This work is integrated with the planning, technical, regulatory, and risk assessment issues covered in the other chapters of this book.

The premise of the report and this chapter is that indirect potable reuse can increase and enhance a community's water supplies. It is best not to propose indirect potable reuse as a solution to a wastewater disposal problem. Asking people to drink recycled water because of a disposal problem is not compelling. As a new and enhanced water supply, however, indirect potable reuse has the following benefits:

1.1 Creates a New Source of High-Quality Water

In some regions, water is transported hundreds of miles, purified, used once by customers, purified again, and discharged into a waterbody. Indirect potable reuse creates a new source of high-quality water by taking advantage of an existing local resource and past investments made in purifying water.

1.2   Improves Drought Resiliency

Being able to offer full water service during droughts is critical for maintaining local economies and quality of life. Most water supplies are affected by yearly variations in rainfall and snow pack, so communities need reservoirs to provide adequate service during droughts. Recycled water is by definition a drought-proof water supply. This means that indirect potable reuse helps increase drought resiliency without having to build costly new reservoirs.

1.3   Maximizes the Value of Water Storage Assets

Many communities currently dispose of treated wastewater even when reservoirs are below optimum levels. Indirect potable reuse returns water to existing reservoirs or groundwater basins, replenishing them even during droughts. This maximizes the value of these storage assets.

1.4   Enhances Water Quality

Because the source water quality is low and because of the negative reaction that most people have to drinking recycled water, indirect potable reuse projects must purify the water to extremely high standards. This typically involves implementing multistep purification processes and rigorous testing. Intensive treatment and testing protocols produce water quality that is superior to that found in most drinking water supplies.

1.5   Is Financially Compelling

Indirect potable reuse leverages previous investments made in transporting and purifying water. It can eliminate or delay the need to build costly new reservoirs. It avoids the cost of new purple-pipe infrastructure because the water is returned to the potable water system and delivered via existing pipes. It uses precious investment dollars to enhance water quality.

Despite these benefits, there are some important issues to address when proposing indirect potable reuse to a community.

2.0   GOALS OF OUTREACH PROGRAMS

In any outreach effort, utilities need to know whom to communicate with and when. Water professionals typically want everything defined and completely designed before starting a dialogue. However, this approach conflicts with some important communication objectives:

- Developing relationships and community support for investment to solve a problem,

- Discussing options for solving the problem,
- Finding people with different points of view who could provide viable alternatives, and
- Understanding people's desires and designing projects that garner *even more investment.*

Trying to *sell* a completely designed project does not embrace the true spirit of the word *communicate*—coming to a common understanding.

It is easy to think that the reason for understanding and managing perceptions of indirect potable reuse is getting the community to adopt it. However, while indirect potable reuse may have compelling benefits, it may not be the best match for every community. This chapter's goal is not to help water utilities *convince* their communities to choose indirect potable reuse but rather to do the following:

- Increase the likelihood that indirect potable reuse will receive fair consideration. Properly managed, indirect potable reuse has compelling benefits for communities that need more water.
- Reduce the likelihood that conflict or organized opposition will
  - Cause the community to make ill-advised investment decisions or underinvest in water quality and reliability,
  - Strand assets because of loss of public support after facilities are constructed,
  - Damage the sponsoring water authority's reputation so future dialogue about water investments is more difficult or strained, or
  - Cause recycled water to be branded negatively or considered generally unacceptable for replenishing potable supplies.

A positive outcome is one in which the community invests appropriately to resolve the stated problem, the sponsoring water utility's reputation is enhanced, and recycled water and its use for replenishing potable supplies is viewed favorably.

The decision to recommend indirect potable reuse tests the utility's basic investment strategies and communication capabilities. Because of the "yuck" factor related to using wastewater for potable uses, utilities must excel at communicating problems, benefits, risks, and water quality. They also must better understand the factors that influence policy decisions.

3.0   IMPORTANT ISSUES TO ADDRESS
3.1   The Problem

Describing the problem to be solved (the need to invest in water supply and water reliability) provides the necessary context for communicating with the community. The magnitude of the problem will determine people's willingness to invest, the amount they will agree to invest, and the level of risk they are willing to accept.

The term *problem* is used in this paragraph and throughout the chapter because people act (invest) when there is a problem to be solved. Without an accepted problem, the result is typically confusion and no action. Utilities need the public to understand that failing to invest in a new water supply will result in a water-reliability *problem* for the community. Also, the type of problem to be solved can significantly affect the dialogue and outcomes. For example, it is a risky proposition to ask people to drink purified wastewater simply because the utility has nowhere to dispose of it.

Because utility personnel typically are technical and solution-oriented, they tend to talk about the solution (the project) before their audience fully appreciates the problem to be solved. As a result, people may claim that utilities are committed to a pet project rather than to solving the problem.

Communications and relationship development efforts must address the *motivations* for taking action or investing. What is the problem? What are all the options for solving it? Without a compelling need and potential solutions, people have difficulty trusting the utility, assessing investment options effectively, and determining acceptable risks. This is why it is critical to begin all communications with a simple description of the important water-reliability issues and the community's specific water-supply needs. Simply put, focusing on *motivations* provides the necessary context for discussing solutions, and brands the utility as committed to solving the problem, not to building a pet project.

3.2   The Source of Water Quality

Several factors contribute to people's perception that the physical source of water is important when determining final water quality. Bottled water companies tout their water sources; for many, the source (a protected spring or other natural source) is an important part of their brand. Water utilities also tend to emphasize the water source in their communications, which further encourages people to connect the water source with final water quality. This is misleading because natural sources of water are rarely potable according to today's health standards. Spring water contaminated with *Cryptosporidium* can be deadly.

Also, you can appreciate why people might be hesitant to embrace indirect potable reuse when you consider their negative gut reaction to the idea of drinking water from a toilet, and concerns that municipal wastewater includes industrial wastes, personal care products, and prescription drugs. This is why building public confidence in water quality is such an important objective. Fortunately, we live in a society that accepts technology's capabilities when managed by a credible organization. We also have the examples of other utilities that became trusted sources of water quality.

It is often frustrating for water-utility executives to respond to concerns about implementing potable reuse because few water sources are free from indirect reuse. Some watersheds already depend on indirect reuse. Mentioning this to the public does not hurt but will not address the trust problem. The issue is whether people trust the utility to manage potable reuse risks and be the source of water quality.

3.3   Conflict

Conflict is stressful, and people tend to avoid it. However, avoiding conflict or attempting to stay "under the radar screen" can be a serious problem when proposing indirect potable reuse. Why? Because this strategy can result in significant and organized conflict after major investments have been made. Then, the sponsoring utility typically is perceived as "committed to the project" because significant dollars have been spent. Conflict that is not addressed very early in the community dialogue can cost millions, probably will tarnish the sponsoring agency's reputation, and will hurt future dialogue about investment. Utility managers need to understand the positive aspects of conflict, find opponents early, and develop relationships with them.

3.4   Politics

The relationship between water utility staff and politicians is important. Elected officials are extremely visible, which means they are exposed to risks that the typical person is not. So, they are very conscious of risk and wary of situations that could damage their reputations and limit future opportunities. Utility staff must understand this concern and design relationship-building and communication efforts that meet the needs of elected officials and policy makers.

In fact, the primary reason for developing communication programs should be ensuring a good policy decision, which means helping politicians minimize political risks. Therefore, the utility must manage conflict and develop a strong foundation of community support for investing in a new water supply and adopting indirect potable reuse. Ensuring a good policy decision is critical because final outcomes will be determined by how safe policy makers feel and how much they trust the utility to keep them out of trouble.

4.0   PERCEPTION, RISK, AND BRANDING

Many would say that indirect potable reuse has been opposed largely because of concerns about water quality and the "yuck factor." While these certainly have been catalysts, there are other important factors. To properly understand and manage perceptions of indirect potable reuse, utilities need to know some basic principles related to perceptions, risk, trust, and branding.

4.1   Risk

Risk is rarely viewed in isolation. People routinely take risks to solve difficult problems or receive benefits. People drive in cars, fly in airplanes, hang glide, and cross the street—all of which involve taking risks to receive benefits. Arguably, any risk is too great if there are no benefits. So, the risks of indirect potable reuse will be viewed in the context of the benefits it creates or the problem that it solves. A wastewater-disposal problem is arguably not the best context for accepting the risks of indirect potable reuse.

Also, taking the audience into account is important when communicating benefits and risks. Benefits will be viewed differently by different individuals or groups. A feature that is compelling to environmentalists, for example, may not be important to other groups or the public.

4.2   Trust

Besides benefits, people also evaluate risks based on whom they are trusting to manage these risks. People routinely trust products or organizations to manage complex or technical issues that they cannot understand themselves. For example, passengers do not request the reliability data on jet engines before boarding a plane. Likewise, most people are not in a position to understand the technical aspects of managing the risks of indirect potable reuse, so they will be looking for an organization to trust. Their trust will depend on the sponsoring utility's behaviors (e.g., motivation for proposing indirect potable reuse and response to opposition) and ability to address water-quality risks.

Trust can be built by starting small and documenting successes before proposing a larger plant. A well-run pilot program that is reviewed by academics and opponents can build trust in the utility and help people accept a proposal to build a full-scale plant. Without trust, indirect potable reuse probably will *not* receive fair consideration, so utilities should make the effort to build public trust.

Private-sector companies build product loyalty and trust via long-term actions. A strong brand is a great asset to have *in place* when a utility begins discussing indirect potable reuse with its community. It makes gaining acceptance of potable reuse easier.

4.3   Brands and Branding

People, products, and organizations are constantly "branded", whether we recognize it or not. They are branded or "labeled" whenever someone has formulated perceptions, made judgments, or developed expectations of them. Negative branding can be devastating, often leading to management shakeups and business failures.

Branding programs use strategies and tactics designed to encourage people to accept or adopt specific perceptions about companies or products. For example, if Volvo wants to be known for building "safe car," they implement a branding program with this goal. Brands are built via publicity, advertising, and direct interaction with a product, service, or organization.

A *brand* is not only a perception but also an asset—what you can "count on" from a product, person, or organization. It establishes a person or organization's credibility to perform certain tasks or provide specific services (e.g., a plumber fixes sinks and toilets). Branding is why wastewater utilities are not credible sponsors of a potable reuse project; they lack the appropriate reputation to manage the related water-quality and health risks. Water utilities are more credible sponsors of such projects.

A widely recognized brand is arguably a business's most important asset. In 2006, The Coca-Cola Company (Atlanta, Georgia) brand was estimated to be worth more than $60 billion. A good brand attracts and retains customers, and is the basis for trust. A bad brand is a liability; it results from negative events or poor management.

A utility proposing indirect potable reuse will be branded on multiple levels, which will affect people's trust and willingness to support the utility. To succeed, the utility must understand which positive brands build trust and will help indirect potable reuse receive fair consideration. One positive brand is the perception that the utility, not the water source, determines water quality. Another is the perception that the utility is collaborative and willing to change based on the public's input. Conducting outreach with no intention of adapting wastes people's time and is the fast track to more conflict and opposition.

5.0   TRUST-BUILDING STRATEGIES FOR WATER UTILITIES

5.1   Treat Audiences as Shareholders and Investors

The local water utility typically is a monopoly, which is an issue that must be considered when leading a dialogue with the community about implementing indirect potable reuse. First, because the utility is a monopoly, customers have no choice, which arguably makes it even more important that they trust the utility. Second, community members are not only customers but also owners (shareholders) and investors. A portion of their water fees funds new infrastructure, so they technically own the water utility's assets.

These ideas are important because water utilities tend to think about the process of communicating with their communities as "public education", which is problematic. Utilities tend to share all sorts of detailed information because they are "educating." Also, utilities may cling to the idea that somehow the public will or wants to understand the technical and scientific issues surrounding indirect potable reuse. Technical information or data will not make the utility's case or build public trust. "Public education" is not the proper focus for sharing project information with key audiences in the community.

When utilities treat audiences like customers *and* investors, they focus on informing them about important assets, meaningful benefits, and needed capital investments. They express costs in meaningful ways—boiling them down to the effect on water rates or fees. Telling our audiences that they need to be educated is condescending and not the path to success. Second, believing that the problem is an uneducated community makes it their problem. It does not encourage the utility to focus on how to *change its behavior* to help indirect potable reuse receive fair consideration.

The following trust-building objectives outline specific *utility* behaviors that will build public trust.

5.2   Invest in Water Reliability

The utility will need to lead a dialogue in which it explains the key factors of water reliability, the local need to invest in new water supplies, the options for developing new supplies, and the compelling benefits of indirect potable reuse. Simply stated, the utility must have something meaningful to say, be willing to listen, and be open to changing course as long as it solves the stated problem. The objectives are for the community and policy makers to trust that the utility understands the important issues, has an open mind about potential solutions, and is committed to maintaining or enhancing water reliability.

The utility's commitment to water reliability is important because opponents have been known to claim that a utility is committed to a pet project or narrowly focused on recycled water. The utility should recommend a solution or a set of solutions, but be open to change. Calling a public meeting and then making it clear that the public's input will not affect the utility's decisions is the fast track to opposition and conflict. On the other hand, if the community invests appropriately in water reliability and the sponsoring utility's reputation is maintained or enhanced, then we can declare victory.

5.3   Create Water-Quality Confidence

The utility will need to build public confidence in water quality by becoming the source of quality, overcoming a common perception that the quality of the physical source is paramount. The community must trust that the utility can manage the real and perceived risks associated with using purified wastewater as a potable water source.

To build confidence and a water-quality track record, utilities should do the following:

5.3.1   Improve the Situation

Make sure that implementing indirect potable reuse will improve overall water quality. People do not like to invest in the status quo and will not be excited about an investment that seems to degrade quality of life.

5.3.2   Use Common Sense

People will be looking for commonsense ways to trust the utility because most know they cannot evaluate it based on a scientific analysis. Utility communications and actions should demonstrate a commitment to increasing knowledge, diligence, and carefulness. For example, testing for more contaminants than the regulations require demonstrates this, as does a water-purification process with multiple barriers and redundant processes. It is important to understand these commonsense connections when planning, designing, operating, and communicating.

5.3.3   Manage Risk

Telling people that "there is nothing to worry about" will not increase confidence. People know indirect potable reuse has risks; they are looking for someone trustworthy to manage those risks. That is why developing trust based on common sense is so important.

5.3.4   Be the Source of Information

People should hear about water-quality issues from the utility before they hear it from others. They should consider the utility to be the expert on water-quality issues. Utilities can demonstrate their expertise by providing third-party advice on selecting home-treatment systems and addressing in-home water-quality issues (beyond the meter).

These four behaviors help build a positive water-quality track record. Negative water-quality events or management problems can undermine this track record, unless utilities respond to them rapidly and effectively, and proactively communicate the situation with the public.

5.4   Turn Conflict and Opposition Into Assets

Conflict (unresolved or fear of) has been the primary reason why indirect potable reuse has been discarded even when its benefits were compelling. Utilities need to find and engage opponents early in the dialogue with the community. Gaining their trust *can* lead to stronger relationships, ardent supporters, and more investment.

One advantage in dealing with opponents is that they are paying attention. Water utilities often have trouble getting people to attend public meetings or be interested in water issues. Disagreement brings attention to issues and could lead to more support and investment if handled productively. Several key behaviors will help the utility manage opposing points of view and conflict:

5.4.1   Lead a Meaningful Community Dialogue

To lead a meaningful dialogue with the community, the utility must select spokespersons who interact well with the public. The utility also must be forthcoming with information to make sound decisions and take into account the ways different people and cultures communicate.

5.4.2   Listen and Embrace Differing Points of View

One major purpose of communicating is to collaborate and collect diverse points of view. This leads to a better understanding of people's needs and desires and potentially to better investment proposals and projects.

It is critical that the community's input affect decisions, or else participants will consider the process a waste of time. This can lead to organized opposition, whether or not indirect potable reuse is part of the dialogue.

5.4.3   Assess the Community With Respect To Conflict

All communities have a history of conflict, which can be a catalyst for conflict in new issues, especially provocative ones like indirect potable reuse. Utilities should identify leaders who were involved in past conflicts and make it a high priority to develop a relationship with them.

5.4.4   Develop Relationships With Opponents

In the short run, the path of least effort is to label opponents "uneducated" or "irrational", but this practice is unproductive and risky. Sincerely pursuing relationships with those who disagree can lead to strong support or at least grudging consent.

The utility should make sure opponents understand the constraints under which the utility believes it is operating. Doing this may lead to a dialogue and solutions that remove the constraints, especially when it comes to investment. Utilities often have capped the investment before communicating with the community. Most consumer research indicates that consumers will pay more if they understand the benefits, and opponents are often advocates for more investment.

5.5 Ensure a Good Policy Decision

Typically, some form of representative government will decide whether to adopt indirect potable reuse, so most of the utility's efforts need to be managed accordingly. Utility managers should gain policy makers' trust by demonstrating that they can manage conflict and build a strong foundation of support among influential individuals and interest groups in the community.

Following are key factors in ensuring a good policy decision.

5.5.1 Understand and Support Policy Makers

Utility managers should assume that policy makers are ethical and want to do the best thing for their community. Claiming to be the victim of self-absorbed or unethical policy makers is neither a strong position nor the way to positive outcomes. Policy makers or elected officials are in the limelight and therefore exposed to higher personal and professional risks. Asking them to take a significant risk for the sake of a water project is not a good strategy. Policy makers are more likely to support a project it if they feel they have the backup necessary to respond to concerned constituents.

Utilities should listen to and understand policy makers' motivations and concerns. They can help support policy makers by cultivating political champions and developing a list of written supporters for a project or investment.

5.5.2 Focus on Developing Specific Relationships

Policy makers must be able to point to a foundation of support when confronted with constituents' concerns. Otherwise, public officials may decide to side with the people who are making the most noise.

Water utilities typically do not have the resources to reach everyone in the community, so they must focus on building relationships with influential people or those who represent organizations or large groups. Key relationships include

- Potential opponents. Anyone who has been involved in past conflicts or identified themselves as an opponent is a high-priority relationship.
- Elected/city officials. City managers, city council members, elected water board officials, county boards, state senators and assemblymen, and congressional representatives in the region *all* fall into this group. These individuals are listened to, represent a large group of people, and may have a political agenda that encourages them to take a position on a controversial subject. Having a positive, interactive working relationship with these decision makers is critical to success.
- Active community members. These individuals are regularly involved in key community issues. They may not be elected officials but are active community members who are listened to by the public or elected officials.

- Business leaders. Because they need a reliable water supply, business leaders are interested in the community's economic condition and position on growth. Many business leaders are also community leaders.
- Ethnic and social group leaders. Ethnic, social, and environmental justice issues can cause conflict when an indirect potable reuse project is proposed. Ethnic group leaders can save or kill a project if environmental justice issue might arise.
- Environmental leaders. Water is an environmental issue, so environmental groups are a relevant and important audience.
- Local regulators. This includes local heath department representatives.
- The media. Representatives of the media often consider themselves independent watchdogs, so developing relationships can be problematic. However, utilities should contact reporters and provide them with important information about the project or issue. A utility spokesperson should be readily available to answer questions.
- Trusted technical or medical community leaders. These groups can be important in developing local water-quality standards and helping people feel confident about water quality.
- Active groups that are recognized. Various groups (e.g., Mothers Against Drunk Driving) have a community voice.
- Well-networked people. Many leaders have a strong network in the community, so they can tell you who else to talk to and where potential opposition may lie.

These individuals can be the nucleus of support or opposition. An ideal collaborator in a public process is a decision maker or representative of a large portion of the community, someone who can disseminate information widely. This collaboration is even more important if an individual has a strong political agenda that relates to past conflict in the community. Project sponsors should identify current conflicts before developing the communication plan. Key individuals will change over time, so the list and communication efforts must be updated constantly.

5.5.3 Communicate With Purpose and Diligence

When utilities communicate, they should listen, learn, develop better projects and methods, and develop support for investment. Utilities need to adopt a collaborative communication style and find out people's motivations and interests. The communication process should document community members' feedback. If someone is a high-priority relationship, the utility should not give up simply because he or she is hard to reach. Utilities may need to hire help to establish a relationship, but diligence and sustained effort are important, and will pay off.

6.0   SUMMARY OF PROJECT CASE STUDIES

The following case studies are evaluated based on the trust-building objectives of investing in water reliability, creating water-quality confidence, managing conflict, and ensuring a good policy decision. This format emphasizes the issues that were significant in determining outcomes.

The insights and advice given here were unavailable when these projects were planned and proposed. They are not a measure of the professionalism or integrity of utility personnel and their consultants. Indirect potable reuse proposals typically require a utility to significantly improve its ability to understand public perceptions, communicate about value and investment, and manage relationships. The people interviewed about these projects were open and forthcoming about both problems and successes.

The following projects are included in this analysis:

- Water Resources Recovery Project (Tampa, Florida),
- Water Repurification Project (San Diego, California),
- Clean Water Revival Project (Dublin San Ramon, California),
- Water Campus (Scottsdale, Arizona),
- Groundwater Replenishment System (Orange County, California), and
- Upper Occoquan Sewage Authority (Virginia).

Other projects are addressed briefly afterward.

6.1   Investing in Water Reliability

When proposing indirect potable reuse projects, success depends on how well the utility defines the water supply problem, presents all the potential solutions, describes the benefits of indirect potable reuse, and demonstrates its commitment to water reliability.

6.1.1   Water Resources Recovery Project (Tampa, Florida)

For years, the Tampa region had been looking for water supply options that would reduce its dependency on groundwater, so water reliability and new water supplies were on the minds of the city and regional leaders. However, two less-controversial options (seawater desalination and a surface water project) and Pinellas County officials' concerns about water quality caused the water resources recovery project to be shelved.

Indirect potable reuse will always be compared to other options, so the benefits and drawbacks of each must be communicated clearly. If all options are not consistently part of the communications, then opponents can claim that the sponsoring utility is committed to a pet project rather than to solving a problem or ensuring water reliability.

6.1.2   Water Repurification Project (San Diego, California)

When it was first proposed in the early 1990s, the San Diego repurification project was the most straightforward, cost-effective way to meet the recycled water goals in San Diego's ocean waiver decision (an allowance by federal regulators to continue discharging primary effluent into the ocean). Although the water was touted as a valuable resource, and recycling was the right thing to do, the city council felt no significant pressure to increase the water supply when they voted down the project in 1998. The drought of the early 1990s was past.

Records show that San Diego's Repurified Water Review Committee considered alternatives, such as nonpotable reuse, and agreed that depending on imported supplies had a number of disadvantages. However, clear, succinct, and quantitatively stated alternatives were not part of ongoing communications. The San Diego County Water Authority (SDCWA) clearly had supported this project, but by the time the city council voted, SDCWA was aggressively pursuing its Imperial County water transfer project instead. The transfer project, which addressed San Diego County water needs, was ultimately approved.

6.1.3   Clean Water Revival Project (Dublin San Ramon, California)

Dublin San Ramon Services District (DSRSD), a wastewater agency, originally sought approval for a new ocean outfall to increase discharge capacity in response to rapid population growth. Dublin's partner communities, Pleasanton and Livermore, refused to support the new outfall. So, the District proposed the Clean Water Revival project to address the disposal capacity issue. The District basically asked the community to drink treated wastewater because it was growing rapidly (faster than both the county and state averages) and could not get approval for disposal capacity. Ingesting something because you cannot find a way to dispose of it is not the most attractive idea. Clean Water Revival was shelved after much conflict and the loss of the local water district's support. Eventually, a new ocean outfall was approved.

6.1.4   Water Campus (Scottsdale, Arizona)

The Scottsdale Water Campus project, which injects recycled water into the ground, was designed to increase wastewater treatment capacity and meet Arizona's Groundwater Management Code. It was successful for many reasons. The community ultimately supported the idea that disposing of a "valuable resource" did not make sense when they needed water to replenish local groundwater.

The alternatives were thoroughly discussed by city officials and a technical advisory committee of academic, regulatory, public health, and technological representatives. A major alternative was to invest in new infrastructure, buy more wastewater capacity from the regional treatment plant, and buy more surface water rights (from the Central Arizona or Salt River projects) to recharge the aquifer. Project communications consistently compared indirect potable reuse with less-attractive alternatives, using simple text and graphics to explain why the indirect potable reuse project was the best option.

6.1.5 *Groundwater Replenishment System (Orange County, California)*

Since the 1970s, the Orange County Water District (OCWD) has injected recycled water into the ground via Water Factory 21 to keep seawater out of the aquifer. Because of consistent communications from OCWD, both professionals and laypersons have come to realize the aquifer's tremendous value and the need to protect it. They know that keeping seawater out solves a major problem. In the Groundwater Replenishment System (GWRS) project, OCWD upgraded Water Factory 21, added more seawater-barrier injection wells, and built a pipeline to transport more water upstream for percolation into the aquifer.

The District communicates that GWRS provides the following benefits:

- Better seawater barrier. Coastal communities know that they must prevent seawater intrusion if they want to continue pumping groundwater for use as a drinking water source. Communities are pumping more water from this aquifer, so a larger seawater barrier is necessary. Using recycled water to maintain the barrier is important because this water comes from a reliable, local source.
- More drought resistance. Storage is necessary for weathering droughts, and a groundwater aquifer is a tremendous storage asset if properly replenished. The ability to inject more water into the aquifer, thereby keeping it at optimum levels, provides more drought resistance. The project will also reduce dependence on imported, climate-dependent supplies.
- No new ocean outfall. Recycling the water allows OCWD to avoid constructing a new ocean outfall. Although estimates vary, the savings have been reported to be as high as $170 million.
- Fewer beach closures. Orange County communities came to understand the staggering economic value of open, pristine beaches when the county led the nation in beach closures in the summer of 1999. The closures attracted significant media attention and led key environmental groups and business leaders to support the GWRS.

6.2   Creating Water-Quality Confidence

The utility must become the *trusted source of quality* rather than simply the physical source of water. This is important because people associate water quality with the water's source (e.g., mountain spring water is healthy). Thinking of wastewater as a source of drinking water is arguably pretty repulsive. Not surprisingly, water-quality concerns were an important factor in all of these case studies.

6.2.1 Water Repurification Project (San Diego, California)

The most visible project sponsor was the City of San Diego Sanitation Department, not the City of San Diego Water Department or the SDCWA. The Sanitation Department managed the North City Reclamation Plant, and the SDCWA had never managed drinking water purification before, so they were less credible than the Water Department. However, the Water Department was not leading the project because of funding or other organizational issues. This was a significant flaw that affected the public's confidence in recycled water quality.

Another problem was the emergence of the "Toilet to Tap" label. It is unclear who actually coined the term, but it was picked up by Steve Peace, a state politician, and repeated by the press to increase interest in an already provocative story. The media received little information countering this connection between the water's source and final quality, and reporters cannot tell both sides of the story if they do not know them. If you want your story told, you must tell it yourself or have other proponents tell it for you.

Water-quality concerns turned into environmental justice issues when people in the communities slated to receive the water felt they were being treated like guinea pigs. The perception was that wastewater was being collected from affluent communities and would be reused mostly in less affluent communities.

6.2.2 Clean Water Revival Project (Dublin San Ramon, California)

This project's sponsor and champion was DSRSD, a wastewater district. Wastewater organizations are not a credible source of drinking water quality, and without a credible water-quality authority in the lead, public confidence in water quality is pretty difficult to achieve. In this case, project opponents felt that DSRSD had no real "water-quality plan" to address such issues as emerging contaminants. The pertinent drinking water authority ultimately pulled its support because of concerns that the project had lost public support.

6.2.3 Water Resources Recovery Project (Tampa, Florida)

In Tampa Bay, even the water professionals were divided on the issue of water quality. One person interviewed stated: "There seems to be a difference between the wastewater and water professionals. The wastewater professionals seem to be comfortable with the water quality, while the water professionals seem to balk at it." This is not surprising because water and wastewater organizations have different requirements and arguably different cultures.

At least one water retailer (Pinellas County) stated that they had "serious concerns" about the variable water quality (especially taste and odor) they already received from Tampa Bay Water. So, it was not surprising that Pinellas County did not support the water resources recovery project. The project ultimately was tabled in favor of less controversial alternatives. This is an example of how perceptions of the existing quality of drinking water (negative or positive) affect people's confidence in and support of indirect potable reuse.

6.2.4   Water Campus (Scottsdale, Arizona)

Scottsdale did several things that encouraged water-quality confidence. Just the name "Water Campus" has a positive, academic image. The city is responsible for drinking water, and they take water-quality issues, including taste and odor, seriously. Scottsdale used a multistep water purification process that includes reverse osmosis and natural purification, and described it clearly via both text and graphics. The Water Campus is an impressive, pristine facility with an on-site laboratory that both conducts comprehensive process analyses and tests water from various points around the city.

6.2.5   Groundwater Replenishment System (Orange County, California)

The Groundwater Replenishment System also is a multistep purification process, but OCWD's water-quality plans, track record, and ethics may be more important. For more than 20 years, OCWD has been injecting recycled water into the ground to create a seawater-intrusion barrier. Staff in its state-certified laboratory consistently test for more contaminants then required by regulations, and they developed the protocols for some of these tests. Also, the District proactively communicates with the public about such compounds as N-nitrosodimethylamine (NDMA) and 1–4 dioxane.

Other confidence-building activities include the following:

- Water-quality studies. The District has been involved in the Santa Ana River Water-Quality and Health Study, which spent between $8 million and $10 million over 6 years addressing water-quality issues. This study is overseen by a dozen or so third-party experts.
- Source water pretreatment. The District supports the idea that the Department of Health Services be responsible for setting standards and permitting treatment processes at sanitation districts because current discharge regulations typically are designed to protect aquatic rather than human health.
- Water-quality ethics. The District has brought water-quality issues to the public's attention even when not required to do so. It also has shut down wells when contaminant concentrations were below the levels at which OCWD had to take any action. This built trust by demonstrating a water-quality ethic that goes beyond meeting regulatory requirements.

So, the Groundwater Replenishment System has virtually no water-quality-based opposition, even though nearby projects have been derailed by the "Toilet to Tap" label. The District's experience, testing, and open communications have made the public, regulators, and health officials confident of the water quality.

6.26   Upper Occoquan Sewage Authority (Virginia)

This project was designed to address significant water-quality problems, including viruses, algae blooms, frequent taste and odor problems, and fish kills, in the Occoquan reservoir. It solved noticeable water-quality problems, had clear and immediate benefits, and clearly improved the situation. The Upper Occoquan Sewage Authority (UOSA) raised public confidence in water quality by creating the Occoquan Watershed Monitoring Laboratory, an independent authority that monitored the watershed's water quality and reported to UOSA and state regulators. This was important because UOSA is a wastewater authority, not a drinking water-quality expert. The bottom line is that people knew the project would improve the water significantly.

6.3   Turning Conflict and Opposition Into Assets

Utilities need to understand the positive aspects of conflict, find opponents early, and develop relationships with those who disagree. Although none of the case studies demonstrate a systematic process of embracing opponents and developing relationships with them, they show that organized opposition is unlikely if your project solves a compelling problem (clearly improving the current situation), if you effectively compare the benefits of indirect potable reuse to those of the other alternatives, and if you develop public confidence in the quality of the water. The projects that were implemented typically avoided conflict because of the nature of the problem and key utility behaviors. The other projects were shelved because of significant conflict or because the community found another solution.

6.3.1   Water Repurification Project (San Diego, California)

Although opposition seemed to "flare up" in 1998, seeds of public discontent can be found in public hearing records during the environmental impact report process in 1996. They were not perceived as serious because, at the time, leading stakeholders generally supported the project. Then the following issues emerged:

- The 1998 National Research Council report on indirect potable reuse—although very favorable—included one phrase that the press quoted widely: that indirect potable reuse should be the "option of last resort" because of lingering concerns about contaminants.
- One journalist at the *San Diego Union Tribune* reportedly never liked the project; proponents claimed that this reporter's opposition inhibited positive coverage of the project.
- A local state politician, Steven Peace, made the project a campaign issue in 1998, positioning himself as "protecting San Diegans" from a "bad project" that would use them as "guinea pigs."

- The project turned into an environmental justice issue. The proposed design treated wastewater that came from some of the more affluent property owners and businesses in La Jolla and then discharged it in the San Vicente Reservoir. This reservoir is predominantly the source of drinking water for lower middle-class Asian, Hispanic, and African-American neighborhoods in west central San Diego. However, it also serves Mission Valley and Hotel Circle, which is an important tourist destination. Despite this, the San Diego City Council representative from the 4th District vehemently opposed the project, saying that an ethnic community had been singled out for an "experimental project."

No information suggests that officials from San Diego or the San Diego County Water Authority reached out and developed relationships with opponents. Nor did they have a base of support (as GWRS did) to help convince the mayor and city council to vote for the project.

6.3.2   Upper Occoquan Sewage Authority (Virginia)

Conflict was generally avoided because the project was a response to a well-documented water-quality problem. It is hard to argue with significantly improving water quality; improvement is a powerful idea. One could argue that the wastewater should have been pumped out of the watershed, but this would have been a major change to the status quo (and customers' experience), amounting to throwing away a resource. Indirect potable reuse had been going on for years; now it was time to improve the water quality.

6.3.3   Water Campus (Scottsdale, Arizona)

This project avoided conflict for several reasons. Wastewater is accepted as a valuable resource in the desert, and the city had built public confidence in its water quality. Also, the project's champion was the water resources manager, who had an excellent relationship with the city council, thereby reducing politically motivated conflict. He also took the editor of the local newspaper to visit Water Factory 21 in Orange County, which was a great way to kick off a positive relationship with the media. Finally, the project was constructed in the growing area of the community, not in the backyard of the antigrowth population.

Ultimately, positive branding, the need for a resource to meet groundwater recharge requirements, and the positive relationship between the water resources department and the city council carried the day.

6.3.4   Clean Water Revival Project (Dublin San Ramon, California)

This project was shelved because of conflict and public perception problems. It was proposed because rapidly growing Dublin needed more wastewater-disposal capacity, not to increase or enhance its water supply. Opponents believed that the project was part of a need to "build out Dublin", whose growth already was considered excessive and detrimental to surrounding communities. Also, the sponsoring utility was a wastewater district, so generating public confidence in the water quality was an uphill battle.

There was conflict between the DSRSD board and the media over headlines. There is no evidence that DSRSD embraced the opposition; rather, opponents believed they were being labeled "uninformed," "no-growthers", and "NIMBY" (not in my backyard). Some opponents felt they were not being heard, and that management was condescending. A small group with a strong scientific base produced a report on why the project was unacceptable. Concerns included the perception that DSRSD had no "water-quality plan" to address emerging contaminants. The District did not respond to the report for 6 months. Opponents also felt that DSRSD was trying to "indoctrinate children" via its school information program. This conflict delayed the approval to start operations. Meanwhile, a new ocean outfall was approved, effectively removing the need for Clean Water Revival.

6.3.5 Groundwater Replenishment System (Orange County, California)

The Orange County Water District conducted an extensive GWRS outreach campaign. District staff avoided quite a bit of conflict, winning over skeptics with their strong message of benefits and innovation, their water-quality leadership and track record, and the professionalism of the outreach staff, who believe strongly in GWRS. Nevertheless, some local water professionals did not think the timing of GWRS made financial sense, and they felt that OCWD staff were not listening to their concerns, but rather repeating their original messages. The District's outreach process makes it difficult for disagreement or conflict to gain traction because OCWD has developed a long list of influential supporters who have put their support in writing. This list was put together before construction began; it gives policy makers the confidence to remain supportive.

6.4   Ensuring a Good Policy Decision

Because the decision to implement indirect potable reuse probably will be made by some form of representative government, a utility must ensure that its communications efforts are focused on *meeting the needs* of policy makers, giving them the confidence to make a decision that results in the best value for the community. So, the utility must integrate the practices of the other trust-building objectives.

6.4.1 Groundwater Replenishment System (Orange County, California)

The Groundwater Replenishment System's outreach program is a good example of what to do when striving for a good policy decision. The Orange County

Water District started communicating with the public many years before construction began and focused on key members of the community. District officials developed a growing list of important individuals and groups willing to provide a signed letter of support for the project. This made policy makers much more comfortable. They demonstrated an ability to win over skeptics with their strong messages, belief in the project, and exceptional track record of water-quality management. The ability to manage conflict is important for making policy makers feel safe supporting any form of potable reuse. The District also has a strong, ongoing relationship with the media, providing them with relevant material for stories.

Another important factor in managing the policy decision is the need for constant communication. District officials communicate regularly with key constituents. The contact list they began building in 1998 has grown to more than 5000 contacts. They add names to the list whenever they have an event or get survey responses. They also update the database when elected officials change. They publish a district newsletter and have created a "Fast Fax" list of more than 250 key people that they use when they need a quick response to negative press. The District also conducts "very important person" (VIP) tours and water classes continually. It has not had a major hiatus in communication since the project started.

Arguably, the Groundwater Replenishment System has an important advantage in ensuring a good policy decision. The District has a governing board of directors, but it serves many Orange County communities that have their own governing bodies. So, the District must reach out to policy makers in many communities, which requires more communications effort. Because it works with so many policy makers, one politician is unlikely to have the power to derail a project. There are too many other important people involved.

6.4.2   Water Campus (Scottsdale, Arizona)

This project's champion was the Scottsdale water resources manager, who had a strong and positive relationship with the city council. He also established a positive working relationship with the media by traveling to Orange County, California's Water Factory 21 with the editor of the local newspaper.

6.4.3   Water Repurification Project (San Diego, California)

Arguably, the San Diego project had flaws that left it vulnerable to political opportunists. It had water-quality confidence issues because the project's champion was the city sanitation department. The early-1990s drought was a distant memory when the vote occurred, the San Diego County Water Authority was arguably more interested in seawater desalination and the Imperial County water transfer, and some city council members did not fully understand the basic merits or value of the project. A broad base of community support had not been developed, and the local paper editorialized against the project.

Political opportunists emerge whenever an opportunity arises. There is little opportunity to use indirect potable reuse as a political lever when there is an accepted need for a new water supply, confidence in the water quality, a strong base of support in the community, and a positive relationship with the media.

The governing structure also affected San Diego's outcomes. It was the city council's decision whether to implement the repurification project, so the decision-making power was concentrated. Smaller governing bodies polarize more easily, and decisions are more likely to be affected by past relationships or issues among council members. So, the environmental justice issue and existing council-member relationships could have been enough to shelve the project. The San Diego County Water Authority was not politically invested in the project when the council voted.

6.4.4   Clean Water Revival Project (Dublin San Ramon, California)

This project addressed the need for more wastewater disposal capacity because of rapid growth. Pleasanton and Livermore, Dublin's ocean-outfall partners, did not support the original proposal to increase outfall capacity, so DSRSD responded with Clean Water Revival, a proposal to pipe reclaimed water to groundwater-spreading grounds in Livermore, with most of the water eventually going to Pleasanton.

Opponents viewed this as a way to build out Dublin at the expense of surrounding communities, who felt that Dublin's rapid growth already was degrading their quality of life. They also viewed it as an environmental justice issue, which typically is guaranteed to cause policy decision problems.

The Zone 7 Water District, the pertinent drinking water authority, initially supported the project. But the District changed its position after a new general manager was hired and a public meeting seemed to show that people did not support the project. The Department of Health Services and the South Bay Water Board continued to approve project operations, but the project was ultimately shelved for two reasons. First, Livermore and Pleasanton approved a new waste-water-discharge outfall, "removing the need" for the project. Second, Pleasanton, the Zone 7 Water District, and a local interest group filed a lawsuit to stop project operations.

To succeed, the project would have needed strong policy-maker support from Livermore, Pleasanton, and Zone 7.

6.5   Insights From Other Case Studies

The following case studies provide more insights on indirect potable reuse. Those involved did not know how much opposition indirect potable reuse could attract, or about the types of communications and utility behaviors needed for constructive community dialogue.

6.5.1 San Gabriel Valley Recharge Project

This project was proposed in the early 1990s in response to a drought, predicted cutbacks, and the need to sustainably manage an adjudicated groundwater basin. The plan was to discharge treated water at five locations into the mostly dry San Gabriel River. The project was shelved primarily because the Miller Brewing Company (Milwaukee, Wisconsin) became so concerned about people's perception of the water used to brew its beer that the company spent significant sums of money opposing the project. This opposition included funding clean water groups, funding candidates who opposed the project, challenging data points, and filing lawsuits.

Anyone implementing an indirect potable reuse project should consider local businesses' branding concerns.

6.5.2 East Valley Project (Los Angeles Department of Water and Power)

When construction began on the East Valley Project in 1997, there was no real appreciation of the conflict that indirect potable reuse could generate. The environment was politically charged because of a mayoral race and San Fernando Valley's proposed secession, which made getting support for indirect potable reuse difficult. It became even more difficult when Jay Leno did a "Toilet to Tap" comedy routine and the Miller Brewing Company threatened to oppose this project as they had the San Gabriel Valley project. The project eventually was discontinued.

6.5.3 Australia (2006–2007)

Australia has sporadic rainfall, and its water system is based on reservoirs. The emphasis on storage makes Australia a good candidate for implementing indirect potable reuse.

In the midst of a multiyear drought, Australian officials attempted to introduce indirect potable reuse in Toowoomba in 2006. The project was subject to a vote because it was connected with federal funding. The community voted "no" after being exposed to significant opposition by a well-known, well-financed community member.

Although no comprehensive case studies have been done, indications are that the failure had less to do with the idea of indirect potable reuse and more to do with how the Toowoomba council conducted the dialogue (people were told that they effectively had no choice) and whom the public trusted. Again, utilities should not underestimate the effect of trust on outcomes.

A subsequent vote planned in Southeast Queensland was eventually cancelled by Queensland's premier, who stated that indirect potable reuse would be implemented if necessary, independent of vote. As of 2007, the story was still playing out.

7.0   WATERSHED DIFFERENCES THAT AFFECT PUBLIC PERCEPTIONS

Differences in watershed conditions can significantly affect how people perceive indirect potable reuse. This may cause the utility to adopt another strategy.

7.1   Dry or Wet Climates

Because the need for indirect potable reuse typically starts with the need to invest in new water supplies, the climate will affect people's perceptions. It is easier to talk about water as a valuable resource if you live in a desert (e.g., Scottsdale, Arizona) than if you live in an area with a lot of precipitation.

This brings up a fundamental issue about how utilities talk about water reliability. Water reliability does not simply depend on whether it rained recently; it also depends on demand, storage capacity, and water-use efficiency. For example, most areas of Florida receive more than 127 cm/a (50 in./yr) of rain, and water shortages still can occur because of demand and storage issues. Utilities must explain their region's water-reliability equation to the public in clear and simple terms. They also need to describe how indirect potable reuse enhances that equation. It's not about the rain; it's about plans, investments, storage, and efficiency.

7.2   Coastal Versus Inland Communities

Differences between inland and coastal communities affect the dialogue about indirect potable reuse. Coastal communities typically have a history of using water once and then discharging treated effluent into the ocean. For them, indirect potable reuse is a significant change and will be perceived as such, which makes the dialogue with the community challenging.

For inland communities, returning treated effluent to the water supply may be the status quo. In Virginia's Occoquan watershed, the issue was not whether to discharge into the water supply, but rather how best to improve the watershed's water quality. In both scenarios, the keys are to stress improved water quality and to make sure the community trusts the utility and believes the improvement is real.

Another difference between coastal and inland communities is the choice of technology for purifying recycled water. For coastal communities, intensive separation technologies (e.g., reverse osmosis) work well because the utility can discharge the brine into the ocean. Brine disposal is more difficult for inland communities, and their water professionals have expressed concern that the public may come to believe that indirect potable reuse is only safe if separation technologies are used. There are alternatives. A well-designed process involving multiple steps, natural treatment systems, and an emphasis on "destroying" rather than "separating" contaminants can give the public confidence in the water quality. The utility must ensure that its processes and communications convey increasing knowledge, diligence, and carefulness.

8.0   TIMELINE FOR IMPLEMENTING AN INDIRECT POTABLE REUSE PROJECT

The following is a timeline for proposing a water-supply project that involves indirect potable reuse. It indicates the tasks needed to build the community's trust, explain the need for new water supplies, and manage water-quality risks. It suggests that 3 to 4 years of work is necessary before the detailed design is created, and that construction may occur in Year 7 or 8.

8.1   Years 1 and 2—Laying a Foundation

In this phase, a utility starts defining its water supply needs and building the trust necessary for indirect potable reuse to get fair consideration. Utilities typically lack the water-quality reputation necessary to overcome concerns about using recycled water to replenish the potable supply. So, the utility should become more aggressive about addressing issues that affect water quality—and avoid events that harm people, the watershed, or the rest of the environment at all costs.

Utility communications should focus on the need for new water supplies and the benefits and drawbacks of the alternatives for increasing supply. The "cost" of alternatives should be expressed in rate changes, which can be estimates.

The utility also should form key alliances and relationships with other pertinent water utilities or agencies (e.g., regional water providers) and key regulators. With past conflicts in mind, utility staff should create a list of desired supporters—individuals and groups—and begin developing relationships with these people.

If possible, this phase should be longer because it is probably the most important part of the effort. Three to four years is better, if the water situation allows.

8.2   Years 3 and 4—Building Relationships and Support

At this point, policy makers have agreed to recommend indirect potable reuse to the community, but a final decision has not been made. The utility should be aggressively developing relationships and constantly assessing both the community's support and the intensity of unresolved conflict. Utility staff should listen to the feedback about investing in new water supply and implementing indirect potable reuse, gathering written support and developing relationships with potential opponents.

Communication materials should address public concerns and utility plans to manage water-quality risks. The goal is to demonstrate the utility's water-quality ethics and diligence, not produce "final" treatment designs. Communication events should be designed to find interested parties or opponents. All of this work will determine whether decision makers agree to proceed with a detailed design of an indirect potable reuse facility.

8.3   Years 5 and 6—Designing and Communicating

In this phase, decision makers should have approved a detailed design of the project. Utility staff should continue to communicate with the community, making the case for indirect potable reuse (including the problem statement and alternatives), gathering and evaluating support, and finding and developing relationships with opponents. The utility should also continue to build public confidence by remaining focused on water-quality improvement and leadership.

8.4   Years 7 and 8—Constructing and Communicating

At this point, policy makers have approved construction of the indirect potable reuse infrastructure. Major capital is being invested, so the financial risks are high. The utility should pay special attention to the people affected by the construction.

Utility staff may be tempted now to declare victory and to reduce communication and relationship development, but these are bad ideas. Communication and relationship development efforts should continue indefinitely to protect the community's investment. Projects have been cancelled after construction was completed—meaning these assets became dormant or underperformed.

9.0   CONCLUSIONS

Although indirect potable reuse clearly can improve communities' water quality and reliability, public support for it will not depend solely on these benefits. It is easy to blame past projects' problems on the "yuck factor", politics, the media, special interests, or "uneducated" opponents because doing so diverts the responsibility from the utility's plans and behaviors. However, if we focus on the utility's behaviors, we can see that developing a community's trust will improve outcomes. This is excellent news because it suggests that utilities have the power and can develop the wherewithal to ensure that indirect potable reuse receives fair consideration.

10.0   REFERENCE

Ruetten, J. (2004) *Best Practices for Developing Indirect Potable Reuse Projects: Phase 1 Report*; WateReuse Foundation: Alexandria, Virginia.

Index

Lightning Source UK Ltd.
Milton Keynes UK
UKOW03f1846120215

246192UK00004B/208/P